Co-Creat[illegible] with Natu[illegible]

"*Co-Creating with Nature* is a call to deepen our relationship with the natural world by moving beyond traditional notions of stewardship. Instead of merely caring for Nature, Montgomery advocates for a profound collaboration—a co-creative partnership with the living ecosystems around us. Filled with personal anecdotes, Indigenous wisdom, and scientific insights, it highlights that Nature is not passive but active, communicative, and willing to engage with us—if only we open ourselves to listening. The practices she outlines, including contemplation, sensory awareness, and sacred play, are designed to heal the separation we have created. This movement from separation to nonduality sets this book apart from other books on ecological consciousness. By collaborating with Nature, we can heal ourselves—a message that feels timely and necessary in our age of separation and crisis."

NISHA J. MANEK, M.D., FACP, AUTHOR OF *BRIDGING SCIENCE AND SPIRIT*

"Pam shows what it means to entertain relationships with other living species. Her story can inspire people who are seeking to do the same."

JEREMY NARBY, ANTHROPOLOGIST AND AUTHOR OF THE COSMIC SERPENT

"A truly remarkable book in every way, indeed, an incredible gift for the world. There is so much wisdom and hope encoded in these pages. There is also scientific research and information to satisfy the most questioning of minds along with beautiful imagery, poems, and quotes to ignite one's imagination. For those searching for a deeper meaning to life and a pathway forward, Pam provides practical guidance and tools. Amidst these chaotic times, it renewed my faith that we are capable of healing our wounds and the wounds of the Earth with the help and guidance of Nature and her healing plants. This beautifully written and utterly important work shows us the way and guides us on the journey."

ROSEMARY GLADSTAR, RENOWNED HERBALIST AND AUTHOR OF *ROSEMARY GLADSTAR'S MEDICINAL HERBS* AND *HERBAL HEALING FOR WOMEN*

"The root of herbal medicine has always been having a deep connection to the natural world and the plants themselves, but in this modern world it can feel increasingly difficult to do that. *Co-Creating with Nature* is a manual to guide us back to remembering that we are a part of Nature as much as Nature is a part of us. This premise shapes our perceptions, morals, pathways to healing, and, ultimately, our practice of herbal medicine."

SAJAH POPHAM, AUTHOR OF *EVOLUTIONARY HERBALISM*

"*Co-Creating with Nature* teaches us how to create a partnership with the plant world, which is so willing to share healing with us if we would just remember our primal connection with plants to guide us in aligning with our true essence. Pam is such a bright light for all of us to tap into. She has an abundance of knowledge, wisdom, passion, and love for working with the plant world in a way that shows us our truth. This book is a true treasure for the times we are living in."

SANDRA INGERMAN, M.A., SHAMANIC TEACHER, AUTHOR OF *THE BOOK OF CEREMONY* AND *SOUL RETRIEVAL*, AND COAUTHOR OF *SPEAKING WITH NATURE*

"Pam's deeply lived wisdom and exemplary guidance in showing us how to enter into an evolutionary and co-creative partnership with Mother Gaia and all her multidimensional and multisentient family fills my heart with love, gratitude, and joy. Remembering our interbeing with all life in a living and essentially loving universe is surely both the greatest wonder and adventure of our lives and the greatest hope for a New Earth to come."

JUDE CURRIVAN, PH.D., COSMOLOGIST, COFOUNDER OF WHOLEWORLD-VIEW, AND AUTHOR OF *THE COSMIC HOLOGRAM*

"A true steward of the plant path, Pam leads us into the wilderness of our own being, opening our senses to two-way communication with the Great Mother. This book is a living bridge, connecting the scientific dots that underlie nature mysticism with such skill that the lines between them dissolve entirely."

NICK POLIZZI, DOCUMENTARY FILMMAKER, FOUNDER OF SACRED SCIENCE, AND AUTHOR OF *THE SACRED SCIENCE*

"Herbalist and Earth Elder Pam Montgomery brings her passionate commitment and deep sense of purpose to shine a light on the far-reaching effects of the primal wound our separation from Nature has caused. She presents a path forward for us to heal ourselves, each other, and our world by learning to listen to and learn from plants, including some of my favorites like white pine and wild rose; to reconnect with ourselves as part of Nature; and to co-create a future that honors and supports all forms of life on our beautiful planet, which she lovingly calls Lady Gaia."

Robin Rose Bennett, herbalist and author of *The Gift of Healing Herbs* and *The Young Green Witch's Guide to Plant Magic*

"Pam's book is prayerful and insightful, filled with a joy, reverence, and depth that come from her years of communicating with the spirits of plants and guiding many students and clients. How Pam weaves an understanding of interbeing throughout this book is truly uplifting and a map needed in these times."

Deb Soule, herbalist and author of *The Healing Garden* and *Healing Herbs for Women*

"Pam takes the lead in sharing cutting-edge information regarding our next steps as fully human partners in our living world. This book lays out a detailed, well thought out journey through the realms of creation many have lost sight of but so desperately wish to reclaim. This helpful and hope-filled book is brimming with insightful connecting narratives *and* practical activities for walking ourselves into healthy and balanced living with the world around us."

Tammi Sweet, cofounder of The Heartstone Center for Earth Essentials and author of *The Wholistic Healing Guide to Cannabis*

"If you are looking for a book to help you remember who you are and how to reconnect with Nature, then this is it! This inspiring and optimistic book eloquently describes how we can come into right relationship with all beings and leave a thriving planet for future generations. Well-informed, thoroughly researched, and packed with heartfelt anecdotes, Pam shares precious teachings from the plants and beautifully describes how plant initiations are a powerful way to build a lifelong loving and co-creative partnership with plants, trees, and the whole of Nature. A masterpiece for our time."

Carole Guyett, medical herbalist, shamanic practitioner, ceremonialist, and author of *Sacred Plant Initiations*

"*Co-Creating with Nature* is not only a profound read but also an accessible and enjoyable one. Pam explores elements within the matrix of Nature (Pachamama) and offers ways to reconnect with her so that we can actively participate in co-creation with other living beings and Nature. These plants have the power to remove our outdated evolutionary and genetic matrix, initiating a quantum and dynamic process capable of triggering epigenetic and regenerative changes at the cellular level. This process aims to help the human species rebuild a natural paradise in harmony with Nature. Everyone should read this book and rediscover their connection with Pachamama."

ROCIO ALARCON, PH.D., HEALER, ETHNOBOTANIST, AND FOUNDER OF IAMOE CENTER

"In this important book, Pam offers a plant-infused healing balm for these uncertain times. With inspirational personal stories and student experiences, she expertly guides us into her deep love of plants and way of being with the more-than-human. Reading this felt like witnessing the gentle light on the horizon just before dawn, a glimmer of hope for our unknown collective future."

RACHEL CORBY, AUTHOR OF *REWILDING & THE ART OF PLANT WHISPERING*

"This book is a powerful reminder of our innate connection to the earth and the importance of a co-creative partnership with Nature. It gently guides us toward living in harmony with the natural world and offers profound insights for those seeking to deepen their relationship with the planet."

IRIS VERBAAS, FOUNDER OF NABALO AND AUTHOR OF *ELEMENTAL WHISPERS*

"Pam's heart is clearly aligned with the soul of the earth. This book is the plant kingdom speaking through her, offering a clear pathway back to right relationship with the self and therefore with Nature. A truly enlightening read."

EMMA FITCHETT (FARRELL), M.SC., FOUNDER OF THE SCHOOL OF NATURAL ESOTERICS AND AUTHOR OF *JOURNEYS WITH PLANT SPIRITS*

"Pam asks us to reconsider our relationship with Nature at a time when it is deeply needed. Her invitation to listen, breathe, and learn from our nonhuman relatives opens a new space for addressing urgent environmental challenges and human loneliness. It is a space filled with wisdom, compassion, and even joy."

STELLA TARNAY, COFOUNDER CAPITAL NATURE AND BIOPHILIC DC

CO-CREATING WITH NATURE

Healing the Wound of Separation

PAM MONTGOMERY

Bear & Company
Rochester, Vermont

Bear & Company
One Park Street
Rochester, Vermont 05767
www.BearandCompanyBooks.com

Bear & Company is a division of Inner Traditions International

Cataloging-in-Publication Data for this title is available from the Library of Congress

ISBN 978-1-59143-522-8 (print)
ISBN 978-1-59143-523-5 (ebook)

Printed and bound in the United States by Lake Book Manufacturing, LLC

10 9 8 7 6 5 4 3 2 1

Text design by Priscilla Baker and layout by Kenleigh Manseau
This book was typeset in Garamond Premier Pro with Montserrat Alternates and Gill Sans MT Pro used as display typefaces

The poem at the end of chapter 2 appears courtesy of Kristin Rothballer.
Line drawings by Lucinda Warner.

To send correspondence to the author of this book, mail a first-class letter to the author c/o Inner Traditions • Bear & Company, One Park Street, Rochester, VT 05767, and we will forward the communication, or contact the author directly at **wakeuptonature.com**.

Contents

FOREWORD

The Impending Quantum Leap

Myra L. Jackson

Humanity is the only living species that resists its own blooming!

MYRA L. JACKSON

We are a people embedded within the myriad systems of this living planet but lack awareness of our interrelatedness to Earth and of the laws that truly govern its varied systems and the systems that animate our human bodies. Much of what keeps our world in motion unfolds from origins that are unseen and remain a great mystery to us.

Although we default to a metanarrative whose logic is often dehumanizing and destructive to the diverse systems that support life, we are born into an interconnected web of life. At deep, intrinsic levels, we know that we are born onto Earth. We know that all life shares the common heritage of all the elements that allow life to thrive. Yet in our rote existence, we are void of feeling a sense of interconnectedness with the planetary being.

We live in the shadowy consequences of our beliefs in separation, our superiority over Nature, and our superiority over others believed to be lesser or too close to Nature to be of value. All life has been harmed as a result.

With my eyes closed, I feel the terror of that lifelong question, *What will it take to break through this limited view and blossom?* This question just now caused me to retreat and fill with doubt, but then I was seized by a distinct feeling of inspired optimism that came from the living Earth herself. The sound of a crow cawing vociferously outside my window latched onto my awareness, and that ever-present bond with nature drew me in close and all of her lessons lifted me. I stopped and withdrew my fingers from the keyboard.

The bird's song reminded me of the Great Mother's prompt to walk every morning at dawn to join the meeting of the birds. During the final weeks of winter in 2011, I found myself out just before dawn, in time to reach a grove of barren trees where the birds were gathering. I stopped each day to listen to their whistling and chatter, and soon stunningly bright verdant leaves began to bud on the trees. The bird chatter simmered. One day, when the leaves were more apparent, I arrived at dawn and all was silent. Tears streamed down my face as I realized that the birds were all sitting quietly as the trees now sang their song.

What a miracle—I witnessed birdsong as a trigger for and announcement of the greening of the trees! It was not long before the trees became the home for a new generation of birds. I witnessed the interdependence, cooperation, reciprocity, and pure love on display among at least seven or more species and six species of trees on this one block in Potomac, Maryland.

Every word to this point appeared in an article I wrote for publication in 2014 called "Living Earth." A much longer article details a clear point of decision I made at the tender age of fourteen to forge a path of return to the Mother. The primordial knower was alive and well in the weave of my DNA and was protecting me from falling into the collective trance based in the illusion of separation. It is relevant to the deep feelings of déjà vu that overshadow me now as I contemplate my experience with the plant Lady's Mantle and the *radical optimism* that courses through what I feel as a precious life.

First, I must speak to the déjà vu. It relates to when I was twenty-eight and working in an investor-owned utility as an engineer. I was also a mother to a remarkable eight-year-old son and actively building upon my experiences of my formative years with devas, sprites, and myriad beings of the subtle world and beyond.

At the time, I was an active member of the Friends of Jung located in Del Mar, California. We met often at a café that hosted known and upcoming authors, researchers, artists, and the like in an intimate setting. In 1986, as the sun hovered above the Pacific Ocean, streaming its saffron rays into the café, a group of thirty of us witnessed a beloved speaker stop twenty minutes into his riveting lecture. Simultaneously lifting his eyes and right arm up to the ceiling, we watched as his hand mimicked a periscope surveying the room. Following this display of strangeness, his now open hand lowered in front of the chair I occupied.

The speaker gazing into my eyes at the time was Terence McKenna. He became a legendary author and experiential researcher of psychotropic plants, psilocybin, DMT, and other novel subjects. He delivered this cryptic message to me:

You. You will never need to ingest the plants that I am speaking of here. You are already connected to the plant teachers.

Without skipping a beat, he returned to his lecture. His voice and the captivating topic swiftly eroded the confusion in the room. Even as I share this, over thirty-five years later, I wonder what happened. It makes no sense, and yet I had received a prophetic vision just two years prior that the way had been prepared for me and that, throughout the days of my life, I would receive messages in surprising and unexpected ways.

For many years, this experience with Terence McKenna was forgotten.

Since 1986, there have been two other times in which this message resurfaced. First, in February 2008, during day two of a thirty-day total darkness retreat in Chiang Mai where I learned that my body was already creating endogenous DMT (while early in the process, this is typical within the pharmacology of the darkroom), when in an instant, the now late Terence McKenna was present in the 360 degree visual theater playing through my third eye. Second, in June 2022, when I ingested the nectars of Lady's Mantle and felt the inner glow of expansion while feeling secure in the embrace of the land and the pulse of the Great Mothers. My exchange with Lady's Mantle was alarmingly clear. She met and matched me where I was and gently elevated my field of awareness and answered a question I had been carrying for a year with great specificity. It was clear that our joining in this way was a part of a larger choreography. We held a shared intention with a date in time and place.

You see, Pam Montgomery introduced me to Lady's Mantle in Vermont at Sweetwater Sanctuary, where a wild water stream that flows there recognizes Pam as family. They live as kin. It's equivalent to being visible in the web of life and available to move with creation, or what sources life to the next greater whole. You can feel the harmonizing forces when humans identify as interrelated, interconnected, and interdependent with Nature. This way of being and doing established an alchemical and biochemical container for the miraculous to rise.

Once in the ceremonial space of this living biome, the intelligence sourcing Lady's Mantle was accessible to all of us. Lady's Mantle met each of us as she found us and helped us all to circulate wisdom whose time has come through each of its own direct links to what sources life. The experience was coherent and congruent, and the cohesion of all present brought us into council with Lady's Mantle and the governing forces of life.

Today, I have opened up into an awareness of the intention that Lady's Mantle holds as a unique facet of Gaia's planetary genus, informing human beings who choose to *be* in relationship with what restores our direct link with what sources life.

Lady's Mantle did provide me with essential keys to the question I had been walking with for a year. The geometry is visible to me now, as is the mathematics of light that takes us from eight to thirteen within the golden mean spiral, which is where we begin to scale in consciousness. It may well be a biomarker for the impending quantum leap.

I am clear that our next evolutionary leap is a quantum one and fully depends upon our relationship with plants and the intelligence they embody. Suffice it to say, this book is relevant for generations of people living now. A legacy book born from a woman who is informed by decades of listening to Nature and applying that earned wisdom into a life way that she can share effectively with others. It is filled with ideas grounded in being and doing in harmony with Nature. In witnessing the brutality of these times, I found myself more optimistic than ever during the months I received the life-affirming chapters that comprise the book before you.

Pam has gifted humanity with a well-honed map for partnering with Nature. The practicalities shine with wisdom culled from a shared collaboration between Pam and the intelligence of the plant relatives she recognizes as true elders. Pam's map is a lamp lighting the bumpy road before the new dawn.

Myra L. Jackson
Germantown, Maryland

Myra L. Jackson is an Earth Elder who has held careers in engineering, holographic organizational development, and academia. She carries the title of Diplomat of the Biosphere with a primary focus on transforming our societal relationship with Nature through public policy approaches that recognize Nature's intrinsic rights to exist whole along with all her life-forms. She also serves on the expert platform of the UN Harmony with Nature Program. Her life's work is anchored by her role as an Evocateur of the Sacred and those ideas whose time has come.

INTRODUCTION

Co-Creative Partnership as a Way of Life

From the time I was a small child I have sensed the aliveness in all that is of Earth and have pursued a deeper understanding of this Earth sentience throughout my life. There have been significant people who have fed my passion for being with Nature, not the least of whom is my maternal grandmother. She instilled in me the simple pleasure of being in true relationship with all the Nature around me, especially the plants and trees. My fondest memories of childhood revolve around the times I spent at my grandparents' farm in the eastern hills of Kentucky. My days were filled with being a garden helper where I would find the little treasures, known as potatoes, in the ground after my granddad forked up the soil. For me it was magical to find these hidden little spuds buried in the earth. Then we would pick and shell peas so my granny could make new peas and potatoes in a creamy sauce. To this day, every year I try to replicate the incredible dish my granny would make with the simple ingredients of peas and potatoes. Somehow my version doesn't taste quite the same. Maybe it's because the magic of my salt-of-the-Earth Grandparents and their living in close relationship with Earth was really the key ingredient. I am deeply grateful for those formative years with my blood-kin grandparents who molded my young heart and spirit into one who cherishes and honors my more-than-human Nature kin.

The first book I wrote over twenty-five years ago is called *Partner Earth*, which is about co-creative partnership with all life. You might wonder why I feel the need to write this book. So much has evolved since that time, and my understanding of co-creative partnership with Nature has grown and deepened more than I could ever imagine. The plight of Nature has also shifted dramatically since then so that climate change has become a climate crisis. I feel an urgency these days to speak far and wide about our need to wake up and step into our rightful place as a part of Nature. The main reason we are in a crisis is because we have fallen into a deep amnesia and have forgotten our birthright of being a part of Nature. The more we separate ourselves from that which sustains us, the more polluted and damaged Earth becomes. Because of our lack of close relationship, Nature has become a commodity to be used and oftentimes abused. We have lost the ability to recognize that Nature is imbued with vast consciousness.

As the need to reclaim our kinship with Nature became more apparent, I founded a group called the Organization of Nature Evolutionaries, known as O.N.E., whose mission extends co-creative partnership further to explore the vast biointelligence of Nature. O.N.E. envisions a future where people become Nature Evolutionaries—becoming co-creative partners with Nature where all life has the right to thrive. We carry out this vision by creating educational opportunities in listening to and building relationships with the living Earth. Honoring our sacred connection with Nature is the foundation of our work.

You will notice throughout this book that the word *Nature* is capitalized. Myra Jackson, a O.N.E. webinar presenter and fellow collaborator, worked diligently to convince the *Oxford English Dictionary* to capitalize Nature. Capitalizing a word signals its uniqueness. As a contributor to the English Language & Usage stack exchange noted: "The Nature with capital 'N' is a divine or ultimate force that controls the universe."

My other creation is an online course called "Co-Creative Partnership with Nature," which I have been running since 2020. Though I prefer being in person with folks when I teach, it was necessary at the time to go online. The beauty of being online is that folks from around the globe have access to the teachings, which has benefited my mission of getting this information out far and wide. Hundreds of students have participated from across the United States and Canada to as far away as Europe, Australia, China, Ukraine, and Cyprus, to mention a few countries. Besides a deep learning experience, the course has provided a community of folks who are becoming Nature Evolutionaries.

Co-creative partnership is just as the words indicate; being an equal partner where the members work together to create and bring about a balanced manifestation. This conscious act of partnering with Nature brings about a union of form and spirit where all life can thrive. Co-creative partnership is not a technique but a way of life that we all carry within us. Embedded in our cells is the memory of a time when we lived close to Earth. During this time we didn't dream of doing harm to that which is the source of our sustenance. But a deep amnesia has taken over, and we have forgotten how to live in co-creative partnership with Earth and all her beings. The separation from Nature that this amnesia has created is resulting in some enormously challenging, changing times. It is imperative that we remember who we truly are and embrace our essence, which is that of being a part of Nature.

If one chooses to embark upon this journey of co-creative partnership, the potential to wake up and remember what it's like to live in harmony with Earth and all her beings becomes possible. A deep sense of well-being is likely to settle into your bones as you experience coming home to yourself and your true Mother, beautiful Lady Gaia.

[illegible]

nto us, as if truly part and parent of us. The [illegible]

not on us but in us. The rivers flow not past, but [illegible]

us, thrilling, tingling, vibrating every fiber and cell of the

ubstance of our bodies, making them glide [illegible]

PART ONE

Moving from Relationship to Co-Creative Partnership

Wonderful how completely everything in wild nature fits into us, as if truly part and parent of us. The sun shines not on us but in us. The rivers flow not past, but through us, thrilling, tingling, vibrating every fiber and cell of the substance of our bodies, making them glide and sing.

JOHN MUIR

ONE
What Is Co-Creative Partnership with Nature?

Study nature, love nature, stay close to nature. It will never fail you.

Frank Lloyd Wright

I have been investigating the vast intelligence within Nature, with an emphasis on plants and trees, for almost four decades now. I realized early on that we humans are deeply connected to Nature and share a symbiotic relationship with the green beings. When occurrences began to happen in my garden that I couldn't explain, I was intrigued and felt further exploration was needed. That was when I took a deep dive into Nature consciousness and glimpsed the magnanimous essence of plants. I entered into a courtship with some of my favorite plants and reveled in the precious communion we developed. These courtships were much more than I could ever have imagined, so much so that these plants were no longer merely friends but became my beloveds. We have truly entered into co-creative partnerships, which have endured all these years.

I want to begin with exploring the nature of Nature. I'm not so interested in the physical laws of Nature but more in the biointelligence or consciousness of Nature. Because Nature encompasses *all* of life, it is crucial to understand what makes up Nature. Many folks think Nature is the outdoors or you find Nature in national parks and that it is outside oneself. The truth is we are a part of Nature, too, so the principles that are inherent in Nature are also inherent in us. Some would say these principles go beyond Earth and are Universal truths. I have contemplated Nature consciousness for almost forty years now, so these principles are not new to me, but it is the first time I am articulating them as a body of knowledge. These same principles are emerging in many different arenas with many different folks from all walks of life at the same time. So, to be clear, these are not *my* ideas necessarily: they come from Nature herself and are part of the vast consciousness of life that we are all part of. Now is the time to cut through the wreckage of modernity, or the "debris field," as Myra Jackson calls it, so that we can wake up to Nature consciousness.

Through morphic resonance thousands of people across the globe are simultaneously experiencing an awakening. Morphic resonance is a concept developed by the renowned biologist and scholar Rupert Sheldrake, who writes: "Each individual both draws upon and contributes to the collective memory of the species. This means that new patterns of behavior can spread more rapidly than would otherwise be possible. . . . Through morphic resonance, the patterns of activity in self-organizing systems are influenced by similar patterns in the past, giving each species and each kind of self-organizing system a collective memory." Morphic resonance suggests that, as a species, we are shifting our reality together across time and space and that when we grow in our conscious awareness as individuals we are contributing to collective transformation.

The Twelve Universal Laws, or the Laws of the Universe, have arisen through time and are thought to be intrinsic, unchanging laws of our universe that ancient cultures have always known about. The laws maintain universal order. Although we may not be consciously aware of the laws, we can confirm their presence through all the phenomena we experience in our lives. The laws are often associated with Ho'oponopono, a meditation for freedom originating in ancient Hawaiian culture, but have also been attributed to ancient Egyptian philosophy. These laws have been expressed by many different peoples in many different cultural and religious traditions through the ages. The author Santi wrote a book about these laws titled *The Twelve Spiritual Laws of the Universe: A Pathway to Ascension*. These universal laws were not invented but rather discovered by humans through observation and experience, in the same way that we have arrived at the physical laws of Nature, such as gravity and relativity. Here is a summary of the Twelve Laws of the Universe, or what I would refer to as the Laws of Nature Consciousness.

1. **Law of Divine Oneness.** The Law of Divine Oneness is the foundational law and states that everything is connected through universal creation. Everything in the universe is an extension of Source energy, and every atom that makes up an individual is connected to all other atoms in the universe.
2. **Law of Vibration.** The Law of Vibration states that everything in this world, whether tangible or intangible, is made up of energy that is vibrating at a specific frequency and that nothing is ever standing still. Further, objects of a similar vibration are attracted to each other.
3. **Law of Correspondence.** The Law of Correspondence states that our external reality is a direct reflection of our internal state and that our lives are created by the patterns we repeat every day. The patterns of our thoughts and actions create the reality around us.
4. **Law of Attraction.** The Law of Attraction states that like attracts like. Our energy field is constantly attracting situations, events, and experiences that directly match what we are sending out to the world.
5. **Law of Inspired Action.** The Law of Inspired Action is related to the Law of Attraction and states when we are aligned with who we are—an extension of Source as stated by the Law of Divine Oneness—our actions become aligned as well, and we take physical steps to move toward our visions and achieve our goals.

6. **Law of Perpetual Transmutation of Energy.** The Law of Perpetual Transmutation of Energy states that energy is constantly evolving or fluctuating. Every thought we have or action we take, big or small, has an impact.
7. **Law of Cause and Effect.** The Law of Cause and Effect states that for every action there is a reaction. The energy, emotions, and actions that we put out to the world will come back to us. In Hindu philosophy, this law is called karma.
8. **Law of Compensation.** The Law of Compensation states that we reap what we sow. What we give shall be given back to us. We will be compensated for our contributions, not necessarily financially but in all the ways the universe can provide abundance.
9. **Law of Relativity.** The Law of Relativity states that reality is a matter of perception. In the universe there is no inherent "good" or "bad"; relativism exists in all things.
10. **Law of Polarity.** The Law of Polarity states that everything in life has an opposite. If there is light there is dark; if there is up there is down; if there is warm there is cold.
11. **Law of Rhythm.** The Law of Rhythm states that cycles are inherent to the universe and everything is in a state of perpetual change. Earth goes through four seasons. We also have seasons to our lives, and all phenomena have a beginning and an ending.
12. **Law of Gender.** The Law of Gender states that the feminine and the masculine are an expression of two primal forces in the universe, and they must be in balance. The masculine principle embodies the active, bright, outwardly focused, and moving aspects. The feminine principle embodies the quiet, contained, dark, and inwardly focused aspects. These two forces are at work both within us and all around us. Historically, the masculine principle has been exaggerated while the feminine principle has been suppressed. Bringing the masculine and feminine principles back into balance is critical to healing our separation from Nature.

Now that we have a better understanding of what I'm referring to when I say Nature with a capital *N*, let's look at the rest of my interpretation of co-creative partnership. When we are expressing ourselves creatively, we allow our imagination to guide us into the infinite possibilities that exist in both the seen and unseen realms. At the root of the word *creative* is *create*. So, action is implied by using our imagination to manifest. Often, this manifestation is original and carries new or unexpected results.

In a partnership we are in cooperation with another. Sometimes we are engaged in an actual contractual alliance where we share a common goal. Within this agreement there are equal rights and responsibilities and no hierarchy between the partners. Together we jointly make a pact to be and act in concert with each other—we *co*-create. Let's look at what Dandelion shared with me to understand what a co-creative partnership looks like in real time.

I had planned a Dandelion plant initiation for the spring of 2020. A plant initiation is a three-day event where you ceremonially ingest the plant and engage with it in various ways (more on this later). When the time rolled around, it was apparent that I would not be able to hold an in-person event because no one was gathering at the time due to the COVID-19 pandemic.

With sadness in my heart I went to Dandelion and explained the situation and said how sorry I was that I was having to cancel the event. Immediately, Dandelion replied, "I'm not on lockdown." I was shocked by this response, and so I asked what Dandelion had in mind. Dandelion proceeded to suggest that I take the initiation online. I had been thinking for a while that maybe I should do some online classes, but I'm technically challenged and never thought I could pull it off, not to mention I could not imagine working with plants via a computer. I told Dandelion I was willing to try but really had no clue how to go about this.

Within a week Kate, a woman who had previously done a Redwood initiation with me in California, contacted me. When I told her about Dandelion, she said she could help and that she did online courses for her university all the time and knew the technical ins and outs of Zoom meetings.

I was amazed how quickly Dandelion had created the environment to begin the process of taking the initiation online. Dandelion and Kate guided me through the entire process, and we managed together to very successfully create a rich and rewarding initiation experience with Dandelion. Dandelion helped me to adapt the initiation to an online format, and Kate helped me learn the technology of it. Attending the initiation was a very skilled online course developer, and she and I went on to create the eight-month online course on co-creative partnership with Nature that I now run. This initiation was the impetus for me to enter a whole new arena teaching about co-creative partnership where there is access for people across the globe. Dandelion carried the big picture, knowing that this collaboration would go way beyond the three-day initiation. I am eternally grateful to Dandelion for partnering with me to help manifest a whole new way for me to bring this powerful work to the world.

The Roots of Separation from Nature

Co-creative partnership with Nature is a conscious joining between humans and any aspect of Nature, be it plants, trees, fungi, soil, water, air, landscape, animals, insects, birds, elemental beings, or nature spirits. Together, we attempt to bring about a balanced manifestation where all life can thrive. Once we begin to engage in co-creative partnership, we initiate the journey of healing the wounds of separation that the all-consuming plague of amnesia has created. Recovery from the trance of separation is the most profound healing we can undertake. This epidemic has created havoc not only within ourselves but also across our beautiful jewel of a planet, Lady Gaia.

There are various historical moments we could pinpoint as the origin of this epidemic of separation from Nature. One place to look is the advent of agriculture twelve thousand years ago, when humans began transitioning away from the nomadic hunter-gatherer lifestyle they had maintained for the previous hundred thousand years. We began to differentiate plants, animals, and land as domestic or wild. We also began storing grain and accumulating personal property. These adopted actions of owning land, sequestering resources, and asserting power over others led us on a trajectory toward the phenomenon of empire building, which began over four thousand years ago with the first known empire of Akkadian, located in what now is known as Syria.

Empire building involves conquest. An emperor, monarch, or powerful government asserts supreme authority over less powerful territories, their peoples and land. In our modern times we could also look at large transnational corporations, which wield immense power as empires. Words like *imperialism, colonialism*, and *globalization* are used to describe empires. The difference between imperialism and colonialism is explained by the educational tutoring app BYJU'S in the following way: "Colonialism is where one country physically exerts complete control over another country and Imperialism is formal or informal economic and political domination of one country over the other. In a nutshell, colonialism can be thought of as the practice of domination and imperialism as an idea behind the practice." *Globalization* is a relatively new term that some see as modern-day colonialism. Ultimately, the building of empires through imperial colonization is a way to control the land and its resources along with its peoples for the benefit of the conqueror. Empires have always seen land (including its water, minerals, soil, animals, trees, and plants) as a resource and have extracted from it to gain wealth and power. Here, we find the roots of our separation from Nature. Once our sovereignty (where we are whole unto ourselves) is taken away by empire-building oligarchs, amnesia sets in. We forget that we are a part of Nature, which includes the land, plants, trees, and water. We forget the language of plants, we forget how to be in cooperation with the animals, we forget what wild water tastes like, we forget how to grow food and keep the soil healthy, we forget what our ancestors knew, and we forget who we are.

Today, there is a movement afoot called decolonization that is attempting to undo the devastating effects of colonialism both to the land and its peoples. This is no small task as empires have been with us and our ancestors for thousands of years and are deeply ingrained in our psyche. So, first we must decolonize our own heart, mind, and body before we can truly reunite with the land and the original people who occupied that land. In my own journey dismantling my "colonizer mind," I discovered a cavernous well of grief.

Colonization doesn't just happen to people; aspects of Nature can be colonized as well. One of the largest colonized parts of Nature is the oceans of the world. Oceans are losing their vitality, and there are even dead zone spots in the ocean where oxygen levels have dropped so much that fish can no longer survive. According to a 2019 article in the *Times-Picayune* (New Orleans), "Dead zones are a worldwide problem. Since the 1950s, more than five hundred sites in coastal waters have exhibited hypoxia, a scientific term for water containing less than 2 parts per million of oxygen. Only about 10 percent of these areas had hypoxia before 1950, according to recent research led by the Smithsonian Environmental Research Center." The article notes that though the dead zone in the Gulf of Mexico is enormous, some 8,776 square miles, the one in the Arabian Sea is seven times larger and is the world's largest. The main contributing factors to dead zones are excess nutrients from agricultural and chemical runoff and rising sea temperatures, both of which cause the water to hold less oxygen.

As if dead zones weren't enough for our life-giving oceans to deal with, huge amounts of plastic are dumped into them as well, which pose the next biggest threat to their health and well-being. While I have known for years that the oceans are being deluged with plastic, when I recently spent time on the coast of Belize, I was shocked by the amount of plastic on the beach, both from people tossing aside their single-use plastics and by what was caught in the seaweed, which offshore currents wash up on the beach. Then I observed a local man meticulously picking the plastic out of the seaweed and loading it in his truck to take home and rinse before putting it on his garden to build the soil. In that moment, as the tears were streaming down my face, a glimmer of hope fluttered in my heart, and I remembered what Dolores Huerta, a civil rights, labor, and women's rights activist said: "every moment is an organizing opportunity, every person a potential activist, every minute a chance to change the world." The next morning I was on the beach with my friends, and we chose to be part of the solution instead of complaining about the plastic. We collected plastic containers and picked plastic out of the seaweed so that it could naturally decompose.

You might say the problem is too big for one person to make a difference, but what else are we to do? Perhaps if we *all* remember our kinship with the ocean, from which all life arose, and treat her with respect, love, and kindness, like we would treat our grandmother, maybe then a new day will dawn.

Decolonizing Language

In working with co-creative partnership, I find that language is so important. As noted earlier, I have capitalized the word *Nature* throughout this book, to indicate that it is worthy of respect. Anishinaabe author, lecturer, and teacher Robin Wall Kimmerer suggests that to stop the Age of Extinction we eliminate the objectification of Nature. Instead of referring to aspects of Nature as *it* we use the word *ki* for singular or *kin* for plural. As Kimmerer says in a 2015 article, "using 'it' absolves us of moral responsibility and opens the door to exploitation." She brings up one more aspect of colonization that we must look at: "Colonization, we know, attempts to replace indigenous cultures with the culture of the settler. One of its tools is linguistic imperialism, or the overwriting of language and names. Among the many examples of linguistic imperialism, perhaps none is more pernicious than the replacement of the language of nature as subject with the language of nature as object."

In the world of herbalism, which I am a part of, I find myself cringing every time I hear how someone "uses" a particular herb. To me this implies taking advantage of the plant, in the negative sense of that expression. My suggestion is to replace the word *use* with *cooperate.*

Open your heart, breathe deeply, and listen to the following sentences.

I'm going to use Dandelion to clear my sluggish liver from toxins in order to restore my liver to health.

Continue to breathe and listen even more deeply.

I'm going to cooperate with Dandelion to clear my liver from toxins in order to restore my liver to health.

How do you feel when you use and how do you feel when you cooperate?

To cooperate implies that you are participating in your own healing process and that you and Dandelion are working in concert. When you use, you are engaging "colonizer mind" where Dandelion becomes a resource or commodity instead of your kin. When you cooperate with Dandelion, you recognize the beginnings of a collaboration that can result in a co-creative partnership together.

Facing the Grief of Separation

One of the more devastating consequences of separation from Nature is grief. This grief is like a mad dog who follows you everywhere, nipping at your heels. If we are not able to turn and face this grief and allow ourselves to feel it, its persistence will eventually occupy every space of our inner and outer being and devour every shred of sanity, hope, and sovereignty. As we are a part of Nature, every cell of our being is sensing the reality of the death, dying, and loss that our species is perpetuating, even if our mind doesn't want to go there. We tend to go into denial so we don't have to experience the excruciating pain. As Stephen Harrod Buhner, the late herbalist, says in his book *Earth Grief*:

> One of the hardest things there is, is to just feel the pain and grief that surrounds us now, to acknowledge it out loud, to be inside the aloneness of it, to not blame anyone for it, to just feel it simply because it is in you to feel it. It's hard but I think it is time to make a different choice, to be braver as a people than we have been, to take the journey that the grief of Earth is asking us to take.

Separation from Nature is at the core of the environmental and existential crises we currently find ourselves in. Instead of denying this grief of separation, is it possible, as my teacher Martín Prechtel suggests, to plant seeds of beauty and hope in the compost of grief that bring new growth?

Author Charles Eisenstein suggests we need to "disrupt the story of separation" and that "the logic of the mind is shifting into knowing that separation is not truth." He urges us to change the story of separation that is "at the core of ecocide" and that one of the ways to do that is by eliminating the enemy because we seemingly "always need an enemy which continually creates separation." Have you noticed how we are always at war with something, like the war on drugs, the war on terrorism, the war on climate change, and so on. Where there is an enemy there most likely is a war. Is it possible to eliminate the warring mentality, thus moving away from the Age of Separation and into, as Eisenstein suggests, the Age of Reunion?

Being in co-creative partnership with Nature is a soothing balm to the trauma wounds of war and thus the inner and outer disconnection. As we slowly heal, a flicker of remembrance lights our way on the path, helping us to find our way home to our true, sustaining, and enduring kindred Mother Earth and Nature. As our partnership grows, we realize that we are actually a part of Nature and that separation is merely an illusion. As the veils of illusion lift, an inherent knowing rises like a phoenix and whispers, "Welcome home."

Nature Trinity

A trinity is a group of three items in a relationship. There are many trinities that have existed for eons and many synchronistic occurrences seem to come in threes. Well-known trinities are: father, son, holy ghost; mother, maiden, crone; sun, moon, earth; life, death, rebirth; body, mind, spirit; brain, heart, and gut—to mention a few.

I became curious about the trinity within co-creative partnership and was inspired by my time visiting Damanhur, an intentional eco-spiritual community in Italy that works with a version of the trinity that Damanhurians refer to as the Three Mother Worlds, which I call the Nature trinity. The three aspects are humans-animals, plants-trees, and nature spirits-elementals. In this trinity humans are participants.

When I contemplate humans and consider what their purpose is, everywhere I look what continues to arise is that humans are here to be responsible stewards for Nature and all life. A steward is a manager who tends to the functioning and well-being of a person, place, or thing. This made me curious about how humanity's role as stewards within the Nature trinity ties into the concept of "managing the forces" that Ecuadorian curandera Rocío Alarcón frequently employs. She suggests that when you manage the forces you engage with the reality around you and focus on the solution, not the problem.

Our role is to manage the forces by actively participating in the manifestation of the living essence of life and Nature. We are the aspect of this trinity that helps manifest spirit into form. Part of our purpose on this planet is to be a steward of the life force because we have the ability to manage in a responsible way. However, we have fallen short in our job of managing the forces in Nature, and many conditions have gotten out of balance because we are not upholding our part of the agreement of participating in the Nature trinity. It is time for us to take up our rightful place within the trinity. We need to look more closely at how we can manage the forces within a co-creative partnership with plants and trees, and especially the nature spirits.

Animals are also a part of the Nature trinity and play an essential role in maintaining a healthy ecosystem. When left to live according to their nature without human interference, they exist in symbiotic and balanced relationships with all of Nature, including all other animals and plants and the soil, air, and water. Unfortunately, at this time there are one million species at risk of extinction, so the stewardship that animals offer is vastly and rapidly diminishing.

Plants and trees are the original members of the Nature Trinity, the elders. They moved onto land 400 to 450 million years ago and hold the long view of life on Earth. They have adapted to be able to survive through hundreds of millions of years of changing environments. They make up approximately 80 percent of all living organisms (when combined with sea plant life), and they are quite successful at creating a life-giving, oxygen-rich environment where all life can thrive.

Trees are stepping up their levels of communication during this fast evolutionary spiral we are in. Their communication travels down through their roots through Earth to communicate, but they are also living antennas who send and receive messages with the cosmos. Part of our evolution is to step into our larger cosmic selves, and the trees are communicating to the cosmos that we are ready to evolve. Because of the communication from the trees, the cosmic beings are recognizing that we are taking up our rightful place as a part of Nature while stepping up to be responsible stewards. The cosmic beings are willing to help us, now, because we are beginning to shift into our true purpose within Nature on this planet.

Within the nature spirit realm, the different kinds of nature spirits could be seen as different races. There are fairies, sylphs, and gnomes, to mention a few. Nature spirits are guardians who help manifest various aspects of Nature. They help manifest balanced life-forms on both a physical and spiritual level. In many areas nature spirits have retreated and even gone into hiding because of humans' abuse and disregard of them. This disregard may partly be because humans are ignorant of their existence. Many believe they are strictly imaginary and not real, and people have relegated them to fairy tales. How can we communicate with them and ask for their guidance in creating a balanced manifestation if we don't even know they exist?

But in some places on the planet, including Iceland and Ireland, humans still believe in nature spirits. They are acknowledged and spoken about openly. Humans who do acknowledge them may designate an area for them to live in and may even build fairy houses specifically for the nature spirits. A fairy house can be as simple as two stones placed in a vertical position with a flat stone on top. This can be just big enough to place gifts inside for the nature spirits, such as beads or bits of food. I place one at the entrance to my sanctuary garden so the nature spirits know they are welcome.

When humans designate a spot for nature spirits to live, this can seem like confining them to a reservation, much like the Native Americans were allotted only certain places to live once the white colonizers took over their land. However, often nature spirits retreat to their own areas where they can live according to their true natures. I understand the reason for this as nature spirits wanting to be free in their expression and perhaps they might feel challenged to participate openly when around humans.

Nature spirits help to bring into reality balanced manifestations of life-forms. Findhorn in Scotland, where fabulous gardens were grown in sandy and originally poor soil, is a prime example of this. The humans and nature spirits at Findhorn worked as a team to bring about beautiful gardens. In the spirit of co-creative partnership and honoring their place within the Nature trinity, we need to ask them to return to our gardens and landscape. We can do this by making fairy houses or feeding them food from our table served on a spirit plate. If we go into an area where they live, we ask permission to be there, and then we give them the utmost respect as they do not look kindly on disrespect. Once you invite the nature spirits back, it is your responsibility to work with them. It would be extremely disrespectful to invite them and then not work with them.

Nature-Human Matrix and Sacred Trust

A matrix is the space from which something else originates, develops, or is contained. Within the context I am speaking about, the Nature-human matrix is the space from which co-creative partnership originates. It holds many aspects of co-creative partnership with Nature. This particular matrix is centered around the relationship between Nature and humans, and because of our symbiotic bond this connection is everlasting. However, this relationship does evolve, and our current evolution is, ideally, from one of Nature as parent and human as child to mature and equal partners.

Implied within this matrix is no separation, or what Thich Nhat Hanh refers to as interbeing. Interbeing is a state of connectedness and interdependence with all life. In his book *The Art of Living,* he writes: "Everything relies on everything else in the cosmos in order to manifest—whether a star, a cloud, a flower, a tree, or you and me." Thich Nhat Hanh emphasizes the interconnection and interdependence of all life even within our own bodies.

> There is a biologist named Lewis Thomas, whose work I appreciate very much. He describes how our human bodies are "shared, rented, and occupied" by countless other tiny organisms, without whom we couldn't "move a muscle, drum a finger, or think a thought." Our body is a community, and the trillions of nonhuman cells in our body are even more numerous than the human cells. Without them, we could not be here at this moment. Without them, we wouldn't be able to think, feel, or speak. There are, he says, no solitary beings. The whole planet is one giant, living, breathing cell, with all its working parts linked in symbiosis.

Even our memories are connected to all of Nature; however, as I mentioned earlier, we are plagued by an amnesia so deep that it has become perpetual. The result of this self-obliteration is that we live within our shadow instead of our light. Herein lies where we begin to separate from ourselves because at an essential level we are biophotonic humans. At the core, the DNA level, of our cells are biophotons, or particles of light, which are connected to the biophotons in all biological life. Through this light we experience our common union and remember our shared journey on this planet. We will discuss biophotons in more depth in the next chapter.

A key ingredient of the Nature-human matrix is sacred trust. In this context we step outside the religious sector and allow our understanding of the sacred to be a power at the core of existence that has a transformative effect on one's life and destiny. When I speak of sacred trust, my understanding is of a transformative power that is reliably true, which becomes a pact or agreement between Nature and humans. When something is sacred, we regard it with reverence and know that it is worthy of our respect. We approach our sacred trust with Nature with honor, grateful for the remembrance that we, too, are a part of Nature. Take a moment to imagine:

> My vision expands as I weave myself into the fabric of Nature creating an exquisite mandala from the warp of love and weft of luminous threads of co-creation. This is what coming home to my original kin feels like. I am thoroughly committed to my partnership with Nature and am ready to make vows for life as I step into my big heart, my Holy Heart. From this one heart space, I sense the miraculous and know that the New Earth is not only possible but already exists.

Making a vow is making a pact or agreement. This is one of the ways we take action to heal our separation from Nature. We will look more closely at this in chapter 3.

Because we are a free-will species we operate by choice. Choice is not a luxury; it is a requirement. What if we choose to be in sacred trust with Nature? What if we choose to live in co-creative partnership with Nature? What if we choose to move from separation into unity consciousness? What if we choose to tell the story of interbeing? Do you love Lady Gaia enough to choose Nature? When we choose co-creative partnership, by the very essence of this relationship, we must consider what Earth and Nature want and not make assumptions. Perhaps if we make this choice and treat Nature as if she is sacred, she will respond as if we are sacred. Taté Walker, a Lakota poet, says, "I want the land to want me back."

Co-Creative Partnership Map

When I'm on a journey I like to know where I'm going, and a map helps me to find my way. I offer you the Co-Creative Partnership (CCP) map on your co-creative partnership journey as a way to plot your course

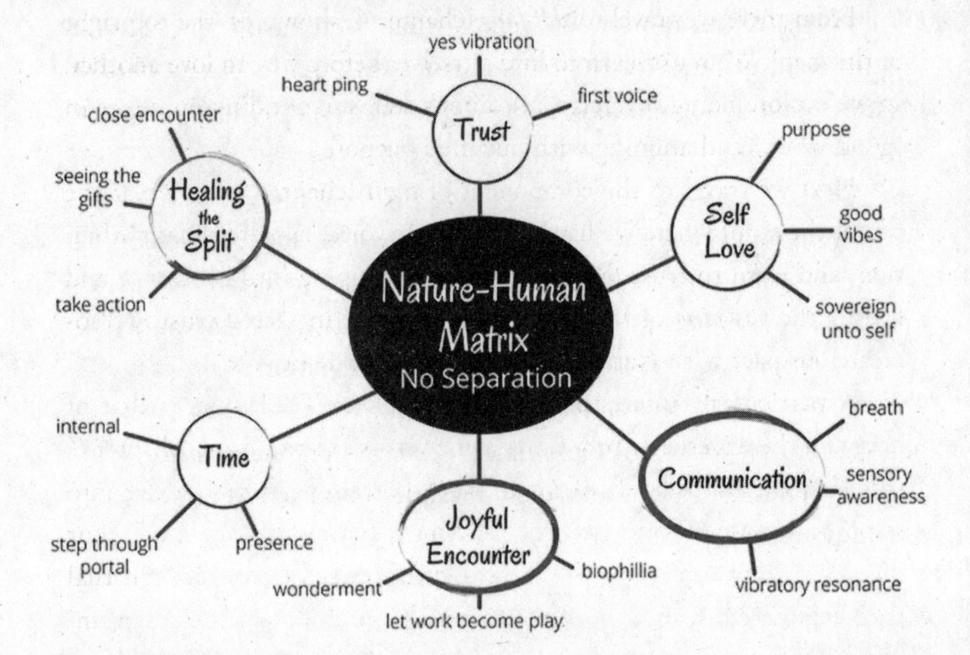

Co-Creative Partnership (CCP) map

and to serve as a guide throughout your exploration. You may find that other avenues emerge to help you find your way into different aspects of partnership. Please add them to the map to inspire future travelers on the path of co-creative partnership with Nature.

We begin our journey in the center with the Nature-human matrix where we explore the original space of separation and nonseparation. We have already discussed this earlier in this chapter. There will be an entire chapter on each of the expressions of co-creative partnership in part two, but I introduce them here.

The first component of the CCP map is *healing the split* (chapter 3, shown on the top left of the map). As we travel through co-creative partnership, the map brings us to the place of separation and how to actually heal the split that has occurred with Nature. In this aspect we see the beauty and gifts in the many forms of Nature. We step into having close encounters as we are guided to take action to heal the split.

From there we travel to *self-love* (chapter 4, shown on the top right of the map). We must learn to love ourselves before we can love another, so we explore being sovereign unto one's self, surrounding ourselves in "good vibes," and aligning with our life's purpose.

Next we travel to the component of *trust* (chapter 5, shown at the top of the map) where we listen to our first voice, explore the yes vibration, and learn to trust our heart through stepping into coherence and feeling the yes *ping* of the heart. As we engage in sacred trust, we co-create our pact with Nature by making commitment vows.

A particularly interesting component of the CCP map is that of *time* and the experience of internal time versus external time (chapter 6, shown at the bottom left of the map). A portal opens as we delve into stepping outside of time and space and move beyond linear time. In our investigation of time the classic expression that originated from spiritual teacher Ram Dass, "be here now," comes alive and we experience becoming completely present in the Now, moment.

The CCP map directs us next to *communication* (chapter 7, shown at the bottom right of the map), one of the most important aspects in co-creative partnership with Nature. Our inquiry takes us into the language of Nature where vibratory resonance opens us to our inherent knowing. We experience our breath as a vehicle of communication. Our senses come alive as we learn about aspects of Nature through sensory awareness.

Finally, we come to *joyful encounter* where we learn about biophilia or the love of Nature (chapter 8, shown at the bottom of the map). Here we can experience wonderment through the WOW moments, otherwise known as Wonder Opens Wide moments. We see that this doesn't have to be hard work but can actually embrace playful partnering as our work becomes play.

This map will serve as our way finder as we embark on this odyssey of co-creative partnership with Nature.

TWO
Coming Home to Nature

The divine communicates to us primarily through the language of the natural world. Not to hear the natural world is not to hear the divine.

THOMAS BERRY

When I think of home what comes into my heart and mind is the familiarity and comfort I feel when I'm home. I have created a space that has an energetic vibration that synchronizes with my own. Home is my sanctuary where safety and sacredness abound. Here I can laugh, cry, love, heal, be inspired, and relax into myself. I am enveloped in the good vibes of my home, and when I am surrounded by good vibes, my energy-body responds and I am in a state of well-being. This is not to say that there aren't times when I don't feel comfortable, like when a closet needs cleaning or a room needs to be reorganized or the plants need to be repotted. But when this occurs, I can feel how off things are so I make the changes necessary to bring my home back to a balance point. When home doesn't make you feel good, then it is up to you to change the environment by either removing what doesn't harmonize or bringing something in that helps bring back the balance.

When I embrace Nature as my home, I feel welcomed into the fold of life-giving creation. I can totally be me when I'm in Nature without façades or shielding. I don't need to be someone else because Nature doesn't judge me at all. As a matter of fact, Nature *wants* me to be all I possibly can be, to shine my brilliance and share my gifts. As with my own home there are times when being in Nature is not comfortable, like when I hear the sound of logged trees slamming to the ground, when I see plastic washing up on an otherwise pristine beach, or when I'm exposed to chemicals being sprayed on an agricultural field.

Descending into amnesia by treating Nature as a commodity to use and oftentimes abuse alters the landscape. This changes the legacy and history of the land. It is suggested that it takes only one generation, or twenty to thirty years, to forget the trees that have been cut down, to forget the plants and their magnanimity, to forget the animals that once thrived, to forget the watercourse that has changed or completely dried up, to forget that there was a spring that gushed water, to forget to tell the stories to our children and grandchildren, and, perhaps most importantly, to forget the songs and prayers that were a part of the landscape and need to be sung and prayed for the land to continue to thrive. Once the legacy or gifts of a landscape are forgotten, the culture of both the people and the land changes. It is the stories of the land that create an unbroken lineage of intactness.

As author Christopher Sandom says in his article in the online news source The Conversation: "Around the world, our unreliable memories and our failure to talk about the natural world between generations means there is an extinction of historical knowledge and experience. This allows important trends in nature to go unnoticed. With each new generation, the current and more degraded state of nature is established as the new 'normal.'" As each generation tells fewer and fewer stories about the land, the amnesia becomes so pervasive that people cease to even be aware that there is a forgotten landscape. The culture of the land is completely obliterated, and the legacy of paradise is forgotten.

Landscapes that have been altered—such as logged areas, polluted waterways, dried-up springs, poisoned soil, heavily sprayed orchards, razed mountaintops, and many others (the list goes on and on)—will recover to a degree, given enough time. But what of us? Since we are a part of Nature we are affected when the landscape is altered. Will we survive these changes? Is it possible for us to take up our rightful place as a part of Nature, being the good stewards we were always meant to be and actually participating in the healing of Nature?

Homecoming Healings

For many years now I have been engaging in what I call homecoming healings. These healings are one small way I have assisted in the healing process when Nature has been violated or damaged. I always listen to my kin (Nature) to learn what is needed and then make suggestions of my own. Together, co-creatively, we determine what is necessary to bring about healing and homeostasis. One of the big ways healing happens is by reintroducing the original blueprint. A blueprint is a complete plan of a design or pattern that guides a physical building or manifestation process. This term comes from architecture where an architect designs a house on paper before any construction begins. When a damaged aspect of Nature is shown its blueprint, it can remember its own originality. Nature's healing is our healing, so when Nature remembers itself we, too, begin to remember who we are.

I have the great good fortune to live in a very beautiful spot with a stream that descends Marble Mountain, goes underground, and emerges right outside my back door. This water is pure, clean, sweet, wild water that, in my humble opinion, is the best water in the world (at least the world I know). This water carries the original blueprint of wild water that has not been polluted or tampered with in any way. When I go to a waterway that I know is compromised, I bring this wild water with me and introduce it into the damaged water. Or I might bring it to a place where a spring has dried up and introduce it to the spring so it can remember when it ran pure and clean and in full vitality.

Another way to help ourselves and Nature return to the memory of home is to give a flower essence to whatever facet of Nature may need it.

A flower essence is a vibrational healing remedy prepared by imprinting spring water with the energetic patterning of a plant and preserving it with a small amount of alcohol or other preservative. I particularly like to give Ash tree essence as it has indicated that it helps to heal Earth and all her beings. Another possibility to help heal Nature is to engage with the Nature spirits and elementals. They play a significant role in tending to the needs of plants, trees, water, soil, mountains, and so on. They engage in maintenance and repair to bring about balanced manifestation. Once we have developed a mutually respectful relationship, we can work in cooperation to reweave the original healthy energetic patterns of Nature.

As we harmonize with the energies of a nature-scape by removing ourselves from artificial frequencies, like computers and cell phones, and sync with the rhythms of Nature, we can move into healing Nature and ourselves. My friend and geomancer Patrick Macmanaway once told me how he cooperated with sound (and the Nature spirits) to bring a dry spring back to life. He sang the spring back to its original brilliance. How is it that we can access the miraculous in this way? As mystic Myra Jackson says, "We can trust Mother Nature and the renaissance she is leading us into. All life can rise whole." And in this wholeness, where vast intelligence exists, miracles happen.

Original Brilliance

When we come home to Nature, we want to bring to Nature our original brilliance, which is contained within all of us. Unfortunately, via the Christian faith, we have been indoctrinated to believe that we come from original sin, which basically means we are born sinners with a proclivity to sinful conduct. Jesus never spoke of original sin, it was developed centuries later. This doctrine was popularized by St. Augustine, a Berber from Northern Africa, who converted to Christianity in 386 CE at the age of thirty-two and published the Confessions of St. Augustine.

Original brilliance is the light each and every one of us is born with. This is not just a metaphor. Each of us contains light that exists at our DNA level in the form of biophotons. Dr. Fritz-Albert Popp, the German biophysicist who built upon the work of Alexander Gurwitsch, was able to prove that biophotons exist. As quoted in a Premier Research Labs blog post, Popp states:

> We know today that man [and woman], essentially, is a being of light. And the modern science of photobiology is presently proving this. We are still on the threshold of fully understanding the complex relationship between light and life, but we can now say emphatically, that the function of our entire metabolism is dependent on light.

For years I have contemplated the phrase "to be enlightened" and have wondered about the origins of this concept. I have found various definitions, each with a different slant, such as "spiritually aware," "free from ignorance and misinformation," and "the state of having knowledge or understanding, self-realization, and awakening." Yet none of these definitions addresses the word *light*, which is at the center of enlightened. The etymology of *enlightened* comes from the old English word *inlihtan*, meaning "to illuminate or to become brighter." With this original understanding, we begin to see that being enlightened has to do with being more brightly illuminated. Is it possible that we are designed to be enlightened by the very quality of our biophotonic essence—our original brilliance?

Our DNA, where biophotons live, is found in the nucleus of the cell. One percent of all of our DNA presents in two intact strands that are in a double helix spiral, and these two strands carry our inherited genes. The rest of the DNA in our cell is noncoded and is not in intact strands but still contains particles of light, our biophotons. There is an assumption that we have the capacity for twelve intact strands of DNA or six double helixes. Dr. Zach Bush, a medical doctor, educator, and thought leader, suggests four groups of three strands. Strands of biophotons act like a beam of light and are often referred to as a "laser" that functions with coherent quantum intelligence and has the ability to communicate with all biophotons in the biological world. Popp, in his published paper "About the Coherence of Biophotons," writes that "it cannot be ruled out that an electromagnetic field of a surprisingly high degree of coherence may be accumulated to such an extent that each molecule in the system is connected (or has the capacity to get connected) to every other one."

What if we could have twelve intact strands of DNA, six times more than we have now? Is this what it means to be fully enlightened? And could our DNA that is not encoded become encoded epigenetically? Epigenetics is the study of how our behaviors and environment can cause changes that affect the way our genes express themselves. This includes our intentions, thoughts, feelings, and communications. If our external and internal experiences (where, how, and with whom) could go into encoding the remaining 99 percent of the DNA in the nucleus of all of our cells, could this play a part in creating our own reality? What if expressing peace, kindness, happiness, gratitude, cooperation, unity consciousness, beauty, friendliness, sacredness, and co-creative partnership with Nature could become encoded into how our species expresses our DNA and passes it onto future generations?

The possibility for us to be all we can dynamically be, living to our full potential as the light beings we are, sounds like a dream I have had my whole life. Is this dream starting to move from the internal landscape of my heart and mind to external reality? If we can dream it then it is not only possible but probable.

Brian Hare and Vanessa Woods, two researchers from Duke University, have written a book titled *Survival of the Friendliest.* Hare and Woods take a deeper look at Darwin's survival of the fittest theory of evolution and suggest that a misguided interpretation of his theory has created a false narrative that stands in the way of us living collectively and cooperatively on this planet. They suggest that Darwin's theory of natural selection, the process through which populations of living organisms adapt and change, has nothing to do with dominance and being the "fittest" but that actually we have survived and thrived as a species through friendliness and cooperation.

Through the study of epigenetics we find that to understand evolution we must focus on how we evolve, the active process. There is a significant difference between these two terms. *Evolution* is a noun and suggests that change is slow moving and beyond our control, whereas *evolve* is a verb that suggests rapid adaptation is possible through our own actions and choices. The Genetic Science Learning Center, which provides tools for teaching genetics, notes that "it takes many generations for a genetic trait to become common in a population. The epigenome, on the other hand, can change rapidly in response to signals from the environment. And epigenetic changes can happen in many individuals at once."

The research of epigenetics suggests that we can evolve in a short amount of time, not over eons. We have the ability to make a massive change in consciousness within one generation. The same amount of time it takes to forget is also the same amount of time it takes to awaken and become the fully enlightened beings that we are designed to be. What path will you take—continuing to forget or awakening? The choice is yours.

Three Centers of Intelligence

There are three areas of intelligence that we can draw upon in our work of evolving—one is centered in the brain and the other two are in the gut and the heart. Each has a unique function. When these three intelligences work in concert, they are referred to as mega brain or whole brain. I prefer to call them intelligences because only one refers to our actual brain, so I think it is confusing to use the term *brain* for the other two. When these intelligences are entrained or synchronized together to a common rhythm, they act as a trifecta and are able to change an energy configuration and create a beautiful coherent synchronicity of original brilliance.

- **Gut intelligence** resides in our belly and guides our instincts. This is where we experience our inherent or primal knowing. We might refer to this as our animal intelligence where we are most connected to Nature.

- **Heart intelligence** is our emotional center and is where we tap into our intuition. There are so many aspects of heart intelligence, such as how the heart understands the concept of infinity and all possible and probable realities as well as understanding union and making the connection with all of life through bonding. The heart can vibrate in harmony and has the depth perception to synchronize frequencies. The heart is able to focus attention on any situation or person or plant; attending allows for no distraction.
- **Brain intelligence** is the logical aspect of ourselves. The brain is able to follow instructions given by the heart and gut and take action to bring the heart and gut's wisdom into being.

The brain generates waves of different frequencies measured in Hertz (Hz), the number of times the frequency repeats in a second. Each wavelength correlates to a different state of being. One state is not necessarily better than another; each has a different function much like the different intelligences. There are five recognized brain frequencies.

Delta 0.5–4 Hz, deep sleep or meditation
Theta 4–7 Hz, dreaming sleep, deeply relaxed, internal focus
Alpha 8–13 Hz, falling asleep, very relaxed, passive attention
Beta 13–30 Hz, normal waking, external attention
Gamma 31–100 Hz, alert concentration

The Power of the Gamma State

I've often wondered what is meant by the phrases "raising our vibration" or "operating at a higher frequency." Then I became familiar with gamma waves. As I investigated gamma waves I found that this particular state enhances our ability to live in the state of our original brilliance. The gamma wave state improves our cognitive abilities and allows for the brain to entrain with the heart so that the heart can function as the primary organ of perception. When the brain is operating in a gamma state, our synapses (the basic form of communication between neurons) are firing more efficiently at a greater rate, and we are in a creative state and able to learn and create new neural pathways and find solutions easily. We become like explorers discovering new territory. The gamma state is also an optimal state for healing—physically, emotionally, and spiritually. Another potent aspect of the gamma state is that we can access unity consciousness—which ultimately means no more war. Could it be that if we catalyze the gamma state we could have peace on earth?

> *Lady's Mantle, whose genus name* Alchemilla *means "little alchemist," was favored by the alchemists. During my initiation with her, she took me deep into her gift of turning the base into gold. As I journeyed with her through the dreamtime and via my focused, activated breathing called Greenbreath, she continued taking me higher and higher into the gold frequency. She revealed how the gold frequency ignites the feminine rising in the numinous Holy Heart, where oneness resides. The Lady weaved her golden thread into my gamma wave state, creating a mantle of unity consciousness where I came home to my Nature kin, and hence to myself.*

Receiving the gifts of Lady's Mantle is one way to access the gamma state. Another way we can access the gamma state is by moving into a state of compassion, a feeling that arises when we understand a person's suffering and feel motivated to help them. Anytime we reach a higher frequency, we move into a gamma state. So, for example, when you have an *aha* moment, the pituitary gland releases the hormone oxytocin, which is produced by the hypothalamus. Oxytocin promotes positive feelings and bonding; it is one of the ways we are hardwired to connect with the self, others, or Nature. During these times of genuine connection when we experience a merging, the high is truly miraculous.

If your desire is to harmonize with the energies of your nature-scape, then the key is to spend time away from artificial frequencies such as cell phones, computers, and televisions or anything emitting super high frequency electromagnetic fields (EMFs). Until about a hundred years ago humans were exposed only to Earth's naturally occurring electromagnetic fields. Even the standard alternating current electricity found all around us can be disruptive. Some antennas used in our increasingly advanced communication systems can emit frequencies as high as 300 GHz (that's 300 billion Hz). Exposure to these frequencies scrambles our natural energetic fields. To bring oneself into a healthy frequency, spend time outdoors (without your cell phone) lying on Earth, being on or near water, walking without a destination through the forest, or sitting in a garden of beautiful flowers.

I find myself next to the ocean, the great Mother Waters where the rhythm of the lapping waves harmonizes with my internal waters as my innate liquid seascape conjoins with flesh and bones. I'm told that 55 to 75 percent of the human body is water, while Earth's surface is about 71 percent water. I feel this mirroring as I allow conscious awareness of my inner water world to merge with this vast ocean, so far from the sweet wild water of my mountain home. I let my heart and mind wander to Sweetwater Sanctuary, the place I call home most of the year, where the purest, sweetest water gushes from a spring emerging from the ground. This water is called structured water because it is always moving, and it resembles the same moving water that surrounds all of our cells. When I drink this water straight from the spring, I can feel all my cells rejoicing in the recognition of water that carries the original water blueprint that nourishes me on the ultimate level of vitality. This life-giving water brings me home to Nature, and I remember who I am.

Symbiotic Relationship with Plants

When we speak of coming home to Nature, we place an emphasis on the part of Nature that is made up of plants and trees because of how closely related to plants we are. They are our ancient ancestors who give us our very life. When we enter the vision of the plants, we remember how to live with Gaia in a sustainable way. We remember our wild heart and indigenous soul—that part of ourselves that never left Nature. The plants tie us to our roots as tribal people who cooperated with one another and the land and lived within the rhythms of Nature.

As John Trudell, the late, great Native American poet, actor, and musician, reminds us: "All human beings are descendants of tribal people who were spiritually alive. Intimately in love with the natural world, children of Mother Earth. When we were tribal people, we knew who we were, and we knew our purpose. This sacred perception of reality remains alive and well in our genetic memory. We carry it inside of us, usually in a dusty box in the mind's attic, but it is accessible."

The Emergence of Plants

Fossil records indicate that plants moved onto land from the sea 450 to 475 million years ago. These first land plants were seedless vascular plants such as Horsetail and ferns. The gymnosperms, plants, and trees who could reproduce themselves by making external seeds (such as conifers who make pine cones) emerged another million years later. One hundred more million years went by before angiosperms, or flowering plants with internal reproduction, became dominant. It was not until flowering plants were well established that humans arrived on the scene. But this is the recent history of plants. The origins go back much further to the dawn of life on planet Earth.

According to Dr. T. Herman Sissons in his book *The Big Bang to Now: All of Time in Six Chunks,* approximately 3 to 4 billion years ago, cyanobacteria (nonchlorophyll microorganisms without a cellular nucleus) began to multiply within the primordial soup of the oceans. These cyanobacteria began using the sun's energy to metabolize water and carbon dioxide, releasing oxygen as a by-product and initiating the process that would become photosynthesis. By 2.3 billion years ago, the photosynthetic process was completely established, and these bacteria had transformed Earth's atmosphere into the oxygen-rich, life-giving biosphere we now live in. The same basic proportions of nitrogen, oxygen, and carbon dioxide that were established then still exist today.

These early algal populations were able to adjust carbon dioxide uptake and oxygen release for millennia to ensure a vibrant biosphere that could support other life-forms, who were also evolving at that time. These biointelligent algal populations were not only able to survive, they also evolved into thriving populations that communicated and organized themselves. Over time, they gained the ability to grow, reproduce, inherit, mutate, and diversify themselves.

A newly emerging field of science known as plant neurobiology is focused on understanding how plants process the information they take in from their environment to develop, prosper, and reproduce optimally. In a paper titled "Plant Neurobiology: An Integrated View of Plant Signaling," published in the peer-reviewed journal *Trends in Plant Science*, researchers suggest that "the behavior plants exhibit is coordinated across the whole organism by some form of integrated signaling, communication and response system. . . . To contend with environmental variability, plants often show considerable plasticity in their developmental and physiological behaviors." Plant neurobiology research puts forward that plants are able to make choices about their evolution, including their transition several million years ago from algal sea dwellers to land plants. Plants alone make up over 80 percent of all living organisms on the planet. This success has continued over the millennia as plants adapt to their changing environment, always moving toward homeostasis. When our human ancestors emerged some 750,000 years ago, plants were well established as *the* stabilizing, life-giving beings on this lovely planet we call Gaia and carried the long view of life on Earth.

History of Human Emergence

To give you an idea of how much older plant life is on this planet compared to human life, I have provided a time line of human evolution. It is difficult to pinpoint our exact emergence, but it has been suggested that the first primates evolved some 85 million years ago, which is 315 to 365 million years after plants moved onto land. Then the first hominid, defined in *Merriam-Webster* as "any of a family of two-footed primate mammals that include human beings together with their extinct ancestors and related forms," emerged 6 to 7 million years ago in Africa. The *Homo* genus appeared approximately 3.3 million years ago. *Homo habilis* is identified as the primary species of this genus during this era and the first to begin using stone tools. *Homo sapiens*, the ancestors of modern humans, surfaced around 550,000 to 750,000 years ago. The word *sapien*, our species' name, means "knowledge." Over time *Homo sapiens* became the dominant species in the *Homo* genus, and eventually all other *Homo* species died out. The subspecies *Homo sapien sapien*, who appeared approximately 160,000 to 90,000 years ago, are modern humans. We are the only species left in the *Homo* genus as Neanderthals, our last surviving *Homo* cousins, disappeared around 30,000 years ago, though Neanderthal DNA still exists in some humans.

Physical Symbiosis

Plants, because they are uniquely photosynthetic, provide us with *all* of our basic physical needs. Through photosynthesis, plants capture sunlight and use this energy to extract carbon dioxide (CO2) from the atmosphere for food. They then combine CO2 with water to form sugars, which make up their roots, stalks, leaves, flowers, and seeds and contain starch, fat, and protein. The by-product of this process is oxygen. Plants, trees, and sea vegetables produce all of the oxygen on the planet. There is no other source.

All of the tissue in our human bodies comes from plants who make their tissues from sunlight. All of the food we eat comes from plants, either directly or from an animal that ate a plant, and all of our oxygen comes from plants. We are completely dependent upon plants for our daily existence. What happens when we become consciously aware that the source of our automatic ability to breathe comes from plants? We realize we are already in a relationship with plants merely because we inhale and exhale. When we become conscious of the source of this breath, our relationship deepens. When we pay attention to the fact that we are exhaling carbon dioxide, which the plants are breathing in, we become aware that we are in a symbiotic reality of constantly exchanging breath with all green beings. We are continually in a cycle of breath, engaged in a relationship of interdependence, even though plants have other sources of carbon dioxide to breathe in.

Cognitive Symbiosis

Edmund Sinnott in his book *Cell and Psyche* postulates that "the rise in higher types of psychological behavior culminating in mind" is a result of plants initiating self-awareness 1.8 billion years ago. Because these ancient plant ancestors were able to communicate with one another and organize themselves to grow, reproduce, inherit, and adapt, they developed self-awareness, which is seen by many in the fields of biological research as the origin of consciousness. During this birthing of consciousness, plants developed sophisticated and complex communication abilities among themselves and eventually with other species, including animals and humans. Plants are able to compute and make decisions about complex aspects of their environment; they have intricate signaling systems for alerting neighboring plants of danger and to exchange nutrients. Within their tissue, plants have large protein molecules that have the ability to store large amounts of information, thus creating an enormous capacity for complex communication and retention of data. Amazingly, they can store memories that guide their future choices. As Jeremy Narby says in his book *The Intelligence of Nature*: "Plants learn, remember, and decide without brains." All of this was occurring millions of years before humans appeared on the planet.

Emotional Symbiosis

Etymologist Edward O. Wilson coined the term *biophilia* to describe our love for Nature, plants, and trees. Because of our long association with plants and our symbiotic relationship with them, we have an inherent (and possibly inherited) need to be in close proximity to them. When we fulfill this need to be near that which gives us life, a type of bond occurs that gives rise to a healthy *gaiacology*—a term used to describe our relationship with a living Earth. Being in close relationship with the source of our sustenance can initiate the release of oxytocin, the bonding hormone, which starts the restorative response. Through this bonding, emotional ties develop, and our hearts are opened to the reality of the mutual love Nature and plants have for us and that we have for them. This is similar to the deep bond between a mother and child. Once the heart is vibrating in resonance with Nature, oxytocin is released and restoration begins. Our inner environment comes into balance, and that affects our outer environment. Not only are we personally restored to a loving balance but we also positively impact all that we come into contact with. From this place of deep nurturance our emotional connection becomes undeniable and we become aware that we are loved and cared for by another being.

Spiritual Symbiosis

Spirit can be defined as the vital principle that gives life or that which animates. Nature carries the strongest vital force. Plants are the most prolific aspect of Nature, and we have always been in close relationship with them. We have direct access to spirit through the plants. The very origins of conscious awareness came from plants, and they have been guiding us throughout our entire species' Earth walk. Each time the plants and trees reach another level in their evolution, humans follow in their wake. It has become evident in the past few decades that plants profoundly inspire humans; their psyche-altering capabilities give people experiences of alternative dimensional realities compared to their everyday lives. Working with psychotropic plants, flower essences, and plant spirits has become increasingly prominent in the world of herbalism and beyond. This tells me that the plants, yet again, are preceding us in our evolution—only this time our evolution is of a spiritual nature. They can guide us, if we only take the time to listen and learn from them.

Plant Initiations

Within the last decade a new way of working with common plants that are not psychoactive has emerged. Plant dieting, the practice of ingesting ceremonially prepared plants, is an initiation that the plants are offering us. In traditional cultures across the globe, adolescents have, for millennia, been initiated into adulthood and formally accepted into society through rites of passage. With a few exceptions, initiations are no longer a part of modern society, so we have become a culture of uninitiated humans who lack maturity and a knowledge of our place within the whole. The elders are gone who remember how to initiate us. The plants are aware that we are longing for the remembrance of connection to the whole, to our people, to our kin—remembering what it means to be truly human within this great web of life. Plants are stepping up and serving as our elders to initiate us so that we can take up our rightful place within the circle of life as co-creative partners with the plants, trees, and elements and Lady Gaia.

In the spring of 2010, a series of volcanic eruptions occurred in Iceland, disrupting air traffic across Europe and affecting some twenty countries. I happened to be in England at the time and found myself unable to fly home to the United States. After many stressful hours back and forth with Mark, my husband, who was at home, we decided it would be best for me to try to get to Ireland to stay with my dear friend Carole Guyett, since we had no idea how long it would be before I could fly home. I made arrangements to take a boat to Ireland, where I gratefully had a warm welcome to stay as long as I needed. This was in the latter part of April approaching Beltane, which falls on May 1 and is the cross-quarter day between spring equinox and summer solstice.

Carole had begun to work with some of her students doing plant diets, which became plant initiations. She was preparing for a Beltane initiation with the Primrose plant. The universe conspired for me to be in Ireland at this time because of a volcanic eruption in Iceland. Wow! This was my first exposure to plant initiations, so I helped her harvest the Primroses and prepare the elixir that was to be ingested. Through working with Carole in her preparations, I began to understand the entire process of placing the initiation within a ceremonial context. I was able to make it home to the United States before Beltane, and Carole sent me home with some of the elixir we had made so that I was able to join in the initiation from afar.

Two years later, in the autumn, the first initiation was held at Sweetwater Sanctuary, my home in Vermont. Carole came to the United States from Ireland to lead this plant initiation. Angelica stepped forward to serve as the initiator for this inaugural initiation.

During the initiation, I sat beside Sweetwater stream where I noticed a lot of bird activity. Birds were coming together as if in a mating dance, but it was September, not mating season, and then they began bathing in the stream. As I was watching this playfulness amongst the winged ones, I became aware of very large wings spread across my field of vision. These wings were quite majestic and I was enthralled by their beauty.

Once the vision faded I went to Carole and asked her about the big wings. She said, "Oh, that would be Archangel Michael." I was a bit surprised and then remembered that the Latin name of Angelica is Angelica archangelica. *It was an eye-opening moment. I said to myself, "Well, of course." I think part of my surprise was that I'm not really an angel person even though I am aware that angels exist. I then became quite interested in Archangel Michael and learned that he is a spiritual warrior who sits at the right hand of God and wields his sword of light in order to separate good from evil.*

When the initiation was over some students were taking pictures and a young woman from California took a photo that captured a large set of purple wings with a streak of light, like a sword, through the middle of the wings. I was stunned when I saw this photo because this was exactly what I had seen during my "big wing" vision (see color plate 1).

Time and time again, prior to, during, and after initiations, miraculous phenomena occur. I have grown to expect them now, even though I'm always amazed by the vastly distinctive ways in which the plants and trees show up or offer unique and unusual components to be a part of the initiation. For example, we held a Rose initiation online, and before the initiation a student of mine who is a fabulous weaver sent me a ball of what seemed like silk because it was so soft. It turns out it was Rose fiber that she had made from the canes of Roses. She sent me enough so that I could send some to all the initiates, who then made bracelets or necklaces out of the Rose fiber so they could wear it during the initiation. This was beyond my wildest imaginings of what the plants are capable of manifesting for initiations. I mean, who knew there was such a thing as Rose fiber?

Now that I have facilitated dozens of initiations, both at home and abroad, what I have found is that the initiations are part of the plant's evolution. They are stepping up to become the elders to initiate us into being truly human and living sanely within the collective. Each initiation is different, but they all offer profound teachings, guidance, healing, and inspiration.

What Are Plants Teaching Us?

Our world is changing rapidly every day, and in a very short amount of time, it will look much different than it does today. Our future becomes one of adaptation to our changing environment. We will not survive with business as usual or by trying to keep things like the old days. We will survive and indeed thrive if we adapt. Plants have been adapting to their environment since day one. They are experts, and now they are teaching us about adaptation. So these times we are living in are not about saving Earth or even saving people; they are about adapting. There are herbs called adaptogens that help us adapt to the many kinds of stressors we are living with. They help us cope with outside influences that the body identifies as nonself and ultimately guide us to engage with what our bodies recognize as self.

Our bodies have evolved for millennia with the natural world, so when we eat wild foods, our cells recognize them as something that optimally nourishes. It is the structure of moving fresh water that surrounds each cell and thus nourishes it. The smell of moist earth in the spring tells our cells that the shift in the cycle is here, and it is time to change our metabolism. Likewise, the sound of geese flying south in the fall signals our cells to shift into wintertime mode. Our bodies recognize the natural world as familiar and supportive, thus causing coherence, aiding in the maintenance and replication of cells. Our bodies do not recognize processed foods; they have to work harder to digest and assimilate any minor nutrients that may be derived from the treated food. Stress is not just the fight-or-flight reaction that produces adrenaline; it is also the added hard work of digesting, filtering, pumping, eliminating, and integrating what is perceived as foreign to our systems. It is extremely detrimental to our overall health to continue to live with stressful life patterns that lead to a lack of coherence, erratic cell signaling, depleted immunity, and low vitality, all of which robs us of the most important nutrient of all—spirit. It is time to look toward our teachers, the plants, to learn about adaptation, resilience, diversity, cooperation, and the vital life-giving principle that sustains us all. The beauty of plants is that they are such magnanimous beings they continually urge one to grow. It's like the old Buddhist saying: "When the student is ready the teacher appears." When you are ready the plants hand you one more piece to help raise your conscious awareness.

My own story of the plants raising my consciousness began in 1969, the year I graduated from high school and met the highly medicinal plant cannabis. In those days what we knew of cannabis's medicine was not about physical healing but about opening the doors of perception so that we could heal on a spiritual level. Throughout my late teens and early twenties, I experienced many plants that taught me about the multidimensional reality of life. Like many in my generation, this was the beginning of knowing deep in my bones that there was another way to be on this planet and that I was a part of this vast biointelligence where a living conversation with the natural and unseen realms is possible. It was not long after this formative time of my life that I learned there were plants called herbs that were equally as potent but in a different way than what I had previously encountered. By the time I was thirty, I was smitten by the plants and trees and wanted to know every little thing I could about them. This love affair has endured for the past forty-plus years, and it only deepens with time. I learned of flower essences early on in my exploration of plants and became intrigued that a plant's vibratory resonance, when transferred into water, could heal the emotional, mental, and spiritual bodies so profoundly. I went on to develop my own line of flower essences while my investigation into the more subtle arenas of healing with plants intensified. By the time I was in my mid-forties, the plants started training me in the vast multiverse of plant spirits. Ten years later I wrote the book *Plant Spirit Healing; A Guide to Working with Plant Consciousness.* Now, as I begin my eighth decade on this sweet Earth, I continue to be amazed how the plants and trees share with me one more bit of their wisdom *when I am ready.* They don't push me beyond my capacity to learn, understand, and integrate new teachings. My love for and commitment to the plants knows no end and I am eternally grateful to be in a co-creative partnership with these magnanimous beings.

Legacy for the Seventh Generation

Native American traditions teach us that our thoughts, actions, and deeds need to always have the seventh generation in mind. What are we leaving for the children? I want to be a worthy ancestor so that my children's children's children will remember and thank me for what I attempted to do for them. My teacher Martín Prechtel reminds us that we are "planting seeds for a time beyond our own" (see color plate 2). What are the seeds that we are planting? Are they ones that support a world where *all* life can thrive?

Turn off your computer.
Put down your phone.
Pick up a feather, a leaf, a bone.
Marvel at a flower.
Plant a seed.
Take a walk.
Remember who you are.
Be generous with your love.
Protect someone.
Take a stand.
Make art.

Dance with strangers.
Find your courage.
Resist distraction.
Be present.
Articulate your dreams.
Act on behalf of the
next seven generations.
Sing to the stars.
Invest in what makes
you come alive.
Open to what is possible.
Decolonize your heart.
Thank your ancestors.
Bless the land.
Remember who we are.

Kristin Rothballer,
The Center for Whole
Communities

PART TWO

Aspects of Co-Creative Partnership with Nature

We need to move: from a spirituality of alienation from the natural world to a spirituality of intimacy with the natural world from a spirituality of the divine as revealed in words to a spirituality of the divine as revealed in the visible world about us.

THOMAS BERRY

THREE
Healing Our Separation

I believe humanity's original wound is that we saw ourselves as separate from Nature.

DR. ZACH BUSH

Healing our separation from Nature is one of the most important undertakings we can attempt at this time. It's as if our internal and external worlds have been split asunder with all the bits being scattered far and wide. I'm reminded of the many dismemberment stories told across the globe and how this narrative of dying to the old and being reborn to the new is now more present than ever but within our modern context. Could it be that we are experiencing a massive, collective dismemberment, and we need help to be put back together again in a new way?

Healing the split involves telling or living a new story where something unique can emerge. My teacher Martín Prechtel speaks of planting seeds where the new plant that grows from the original seed looks very different than it did in the past. It carries the original blueprint within the seed and yet what emerges is an amalgam that has been shaped by a new set of energetic circumstances. He says, "I'm not trying to reinstate what people have had and lost. I'm trying to take something from the past and make it come back alive again in the seed germ of what is still here. There are intrinsic and innate things in all people where the actual spiritual DNA knows how to remake itself, not in the form it was, but in an adjusted form within the conditions it finds itself." In other words, we take the wisdom of the original seed (our spiritual DNA) and plant it in new conditions, new soil, or a new milieu and let it grow into what it will, according to its own true nature, which will not be the same as the original seed and what sprouted from it. Then we don't judge the new plant but honor the fact that it is alive and growing a new story. When we take the brilliance of spiritual DNA and remove it from the religious arena, a whole new vision of our potential is illuminated.

I'm going to travel down a "what if" path that some of you may think is idle fantasy, but perhaps for a moment you could put aside your doubts and journey with me.

If we believe that the macrocosm is reflected in the microcosm and the dismemberment that is occurring externally has also occurred internally in each human and all but two intact strands of DNA have been split asunder—dismembered—thus creating noncoded DNA that is not intact, then does that mean it can be put back together again in a never-before-seen way? Through epigenetics do we have the ability to encode the bits of DNA with peace, love, cooperation, and unity and live what has only been a dream, but one we feel to our core is possible? Through engaging in co-creative partnership with Nature and unity consciousness, can we "re-member" all the bits of DNA creating intact strands filled with light, thus becoming enlightened while living a new story, one we can hardly even talk about because we have never seen it or lived it? Is this what it means to remember? To quote a beloved prophet of our times, John Lennon, "You may say I'm a dreamer, but I'm not the only one. I hope someday you'll join us, and the world will live as one." If we can dream it, it is possible, so let's dream big.

Discovering Nature's Gifts

Every single aspect of Nature carries gifts, beauty, and purpose. We all are here for a reason and have our unique path to follow, living according to our own true essential nature. There are times when we perceive a certain aspect of Nature to be an enemy, such as ticks who carry Lyme disease, an invasive species of plant, animals who eat our garden, or polluted water that is undrinkable, to mention a few. One of our big challenges is to see the gifts in "the enemy" and to partner co-creatively with that which we fear will do us harm.

We begin healing our separation from Nature by discovering what these gifts are and asking our so-called enemy: "What is your purpose on the planet? Why are you here?"

My student Elizabeth shares her story of partnering with slugs.

We have had one atmospheric river after another in the Pacific Northwest. Moss is growing everywhere, *even out of the bark of my weakest birch tree! The soil is beginning to smell, and worst of all, slugs have been sharing my spring lettuce. So yesterday I sat with the beingness of slugs. I was shown their role as Mother's composters. How they move creates sacred geometry patterns, and their slime contains sacred geometry potential for the nature spirits to play in. I met their king and explained that I need my spring lettuce. I don't kill the slugs, but I do throw them into a nearby wild patch of greenery with a bit of anger. He asked me to pick them up, honor them, and place them gently on a leaf in the wild green patch, so I did. I told King Slug I was putting beer traps in my lettuce patch, but I hoped he would keep slugs away for their safety. I honor their role in Mother's design but am not willing to give up homegrown spring lettuce.*

This morning, for the first time, no slugs. We'll see if it lasts. What will last is my honor and respect for what they do. King Slug tells me this is a lot, since most humans are repelled by them. All honor and respect to the slugs!

Through her intuitive sense, Elizabeth discovered the gifts of slugs, both what they do for the ecosystem and the beauty they create with their geometric patterns. She took action to treat them with more loving-kindness, which is what they wanted and needed.

In this next story, my student Martha shares her experience with yellowjackets.

Earlier in the year, I noticed a yellowjacket nest in our shade garden, about ten feet away from the front porch, where we often eat dinner in the summer. I sighed, recalling how aggressive these little creatures can be, especially when food shows up in their vicinity. I purchased a can of the essential-oil-based wasp killer I've used in the past.

Then I remembered our course. I had a talk with the yellowjackets. I told them I didn't want to hurt them or their home, and I would leave them alone if they wouldn't bother us while we were eating or chase us around the yard. We had quite a few outdoor dinners and not a single incident of yellowjackets buzzing our plates all summer. I've gardened all around their nest, leaving them a four-foot radius that I do not enter. None of them have chased me. Wow!

Martha's story demonstrates making an agreement with the yellowjackets that was mutually beneficial.

Many years ago (about twenty-eight), I had an experience with Poison Ivy that set the precedent for my healing of separation from Nature. I have told the story on the next page many times over the years because it is such a great example of how we can move into a co-creative partnership with Nature. This story originally appeared in my first book *Partner Earth: A Spiritual Ecology.*

I was presented with an opportunity to heal a very big split one winter when I had a Poison Ivy rash all over the side of my face. As a kid I was often covered with Poison Ivy rashes. I walked around coated in pink calamine lotion. Poison Ivy and I just couldn't seem to get along. A few winters ago we were burning wood that had Poison Ivy vines on it. It never occurred to me that I could get a Poison Ivy rash in the winter, so when a horrendous rash appeared on the side of my face, I had no idea that it was Poison Ivy. I kept putting salves on it, and it just kept spreading. It progressed to the point where the whole side of my face was covered in crusty oozing blisters. Since it was on my face, I was "faced" with it every morning in an inescapable way. One day, at my daughter's school, a little girl came up to me and said, "How come you have pizza all over your face?" At that, I decided I needed to resolve this problem.

I went home and stared at myself in the mirror. In a flash I realized that this was Poison Ivy, and I vowed at that moment that I would change my relationship with the plant. That spring I observed Poison Ivy closely and listened to what the plant had to say. I began to recognize the gift that Poison Ivy has to offer us. It grows mostly where the land has been abused at some time. In my case, I live where there used to be orchards that were heavily sprayed with chemicals. Poison Ivy is there to protect the land and help it recover from the trauma that was caused by the application of will.

I began to have a profound respect for Poison Ivy and an understanding of the great service it gives to the land. The reality, of course, is that I still need to be in and among Poison Ivy because I harvest herbs in spots where it grows. I began that spring to eat three tiny red leaves once a day for five days. This helped my body match its resonance with that of Poison Ivy, and now I may only get a trace of the rash on my wrists. Every spring, as I eat my three red leaves for five days, I thank Poison Ivy for the wondrous gift it gives to us all, not just to the land, for when the land heals, we heal. This experience was my way of healing my split with Poison Ivy.

I went on to make an agreement with Poison Ivy where I agreed to be the protector of the land so that Poison Ivy wouldn't need to do that job anymore. In exchange Poison Ivy agreed to recede from my gardens.

*I began to fiercely protect the land, which was a commercial farm, and not let any Nature kin be killed or abused in any way. In accordance with the pact we had made, each year Poison Ivy receded a bit more from my garden until it only lived on the far edges of the property. I moved away from that area but since that time, over a quarter of a century ago, I have never gotten a Poison Ivy rash again. (*Note: *I am in no way advocating for you to eat Poison Ivy. This was* my *agreement with Poison Ivy.)*

Plant Gifts

The separation from plants and trees has been massive despite a green revolution, which began in the 1970s, especially in the arena of medicinal herbs. Trees are mostly seen as a commodity, and plants are seen as weeds that need to be eradicated from agricultural lands and suburban lawns. There is a relatively new term, *plant blindness*, also called *plant awareness disparity* (PAD). In a 1999 article, U.S. botanists James H. Wandersee and Elisabeth E. Schussler defined plant blindness as "(a) the inability to see or notice the plants in one's own environment; (b) the inability to recognize the importance of plants in the biosphere and in human affairs; (c) the inability to appreciate the aesthetic and unique biological features of the life forms that belong to the Plant Kingdom; and (d) the misguided, anthropocentric ranking of plants as inferior to animals, leading to the erroneous conclusion that they are unworthy of human consideration." In addition they noted that "plant blindness is a factor in the ongoing declines in university botany programs, herbaria, and other plant science facilities."

Another way that separation occurs with plants is through the diminishing number of plants that are available to us as food, called plant reductionism. On average there are about twenty species of vegetables and fruits available to us at the grocery store—and these are in the more well-stocked and diverse stores. These twenty species make up about 90 percent of our vegetable and fruit diet. However, there are twenty thousand species of edible plants across the globe. Why are we so limited in the number of plants available to us? Is this contributing to our loss of vitality and overall lack of health?

My neighbor and regenerative farmer Ryan Yoder from Yoder Farms has this to say about plant reductionism.

> *The overall activity to bring us back to health is what will bring the ecology of the planet back to health, since we and our biomes are extensions of earth and her biomes, really complicated on the face of it, but actually the steps needed are simple and clear—regenerate life, heal the soil, end chemical warfare on people and the planet, rebuild biological diversity of the soil and ecology and focus on causal factors rather than treating symptoms of the problems. Start at the grass roots and recognize that it begins with each of us applying our agency and capacity and insight to make changes where we are!*

When we look at the gifts of plants, there are so many. I'm reminded of the poem by Elizabeth Barrett Browning (Sonnet 43). She may have been speaking of a person, but for me I feel this way about plants. I share Barrett Browning's poem to give you a taste of my profound love for the green beings.

How do I love thee? Let me count the ways.
I love thee to the depth and breadth and height
My soul can reach, when feeling out of sight
For the ends of being and ideal grace.
I love thee to the level of every day's
Most quiet need, by sun and candle-light.
I love thee freely, as men strive for right.
I love thee purely, as they turn from praise.
I love thee with the passion put to use
In my old griefs, and with my childhood's faith.
I love thee with a love I seemed to lose
With my lost saints. I love thee with the breath,
Smiles, tears, of all my life; and, if God choose,
I shall but love thee better after death.

While sitting on my wintertime porch in Belize, with the palms rustling in the breeze and the intoxicating scent of Plumeria wafting through the air, I contemplate the gifts of plants and trees. Let me count the ways.

> Plants carry the gifts of life-giving oxygen while providing us with food that sustains us and forms our tissue; plants have fiber that can be made into clothing or cordage; plants provide medicine for every possible disease we encounter, including emotional and spiritual healing that is needed; plants open the doors of perception; plants share a symbiotic partnership with us; plants provide guidance and teach us how to adapt; plants have the long view; plants bring joy because they are beautiful; plants communicate with all life; plants support bees and other pollinators and provide sanctuary for insects; plants are protectors; plants nourish soil by fixing nitrogen and adding humus; plants are light transformers.
>
> Trees provide materials to build shelters; trees carry the gift of fire inside their core helping to provide warmth, the ability to cook, a place in which to tell the stories of our ancestors, and a central focus for ceremonies; trees are majestic and inspire awe; trees provide shade and a home for birds and other animals.

When I take in this list, which is by no means complete, of the true essential nature of plants and trees, I can hardly imagine how plant blindness or, the more appropriate term, plant awareness disparity is even possible.

Water Gifts

Alongside soil, water is a critically important aspect of Nature, and it is crucial for us to heal our separation from it. Myra Jackson, who is a representative member for the Harmony of Nature committee at the United Nations General Assembly, says, "When the oceans die, human life dies."

I wrote these words and became overwhelmed with a wrenching grief, so I put my shoes on and walked the two blocks to the Caribbean sea.

The Mother Waters have a churning surf this morning as a big blow came through yesterday and persists today. I watch the huge orange ball of sun rising out of the water, and I give thanks for the dawning of a new day, welcoming what is to come. I raise my hands into prayer pose and begin to speak with my beloved Caribbean sea that, over the last four decades, I have come to know intimately in all her guises (see color plate 3).

Oh, Great Mother, I come to you this morning with heaviness in my heart and want to say I'm sorry. I'm sorry that humans have stripped you of so many of your dwellers. I'm sorry you are being choked with plastic. I'm sorry your coral reefs are being bleached. I'm sorry your temperature is rising, causing all manner of chaos. I'm sorry your whales and dolphins are being deafened by sonar waves. I'm sorry you have been so disregarded and abused.

I ask for forgiveness for me and my kind—even though we don't deserve it. When you sing to me with your undulating rhythms, I will listen. I hear you, I see you, I am in service to you.

You, Thalassa, primordial goddess of the sea, are the originator of *all* creation. You carry the memory of primal beginnings. You, with tidal force, can create anew.

Water carries so many gifts, but one of the biggest is that of holding memory. This is the main reason water is used as a medium for making flower essences. The water receives and records the vibratory imprint of the flower and then transmits this imprint to the receiver.

The German scientists Dr. Bernd Kröplin and Regine Henschel conducted experiments on water's ability to store memory. In his book *The World in a Drop,* Kröplin demonstrates how introducing vibrations to water, like those of flowers or crystals, changes its structure. Water also takes on the vibratory resonance or energy of the environments that it travels through and then stores the memory. Kröplin went on to show that humans are affected by the vibrations around them because the 70 percent of water we are made up of holds the memory of our internal thoughts and feelings and external terrain. So this means that *everything* we take in and surround ourselves with affects the water that surrounds and is enveloped within our cells.

In *The Hidden Messages in Water*, the well-known work by best-selling writer and researcher Masaru Emoto, the author demonstrates how our thoughts and emotions can change the structure of water. He exposed water to both negative and positive thoughts and emotions and then froze the water and photographed the crystals that formed. The difference between the two was quite dramatic. He also wrote down words and phrases with various meanings, such as *love* and *I hate you*, and sat a glass of water on each word. Even this changed the structure of the water. Emoto organized a form of "homecoming healings" by holding prayer circles at polluted water sites to help change the structure of the water. His amazing photos can be seen in his book.

In my own experience I certainly can tell the difference between drinking the wild water from Heart Spring at Sweetwater Sanctuary, the place we call home in Vermont, and drinking water that is not from such a pure source. I often refer to myself as a water snob because I have a hard time drinking water that is not of good quality. The bar is set fairly high, though, because I consider my water the very best you can drink anywhere in the world.

My mentor and friend Rocío Alarcón, who is an Ecuadorian curandera and ethnobotanist, was visiting Sweetwater Sanctuary in Vermont a few years ago to facilitate a workshop in healing and learning the ways of the shaman. During this workshop a profound experience happened.

Rocío wanted to share the practice of spiritual bathing in our stream at night. So my husband, Mark, set out up the mountain to find an appropriate spot. He scouted up and down the stream, taking into account all of the requirements. The spot needed to be relatively close to the teaching center but still secluded and not have too many obstacles so folks wouldn't stumble or fall on their way there. It needed to be a narrow enough section of the stream so that crossing would not be too perilous, and there needed to be enough open space on the banks for several folks to fit. Mark found the perfect spot where we could hold our nighttime ritual bathing ceremony, but, of course, we had to show this spot to Rocío for her approval.

The morning of our planned bathing ceremony we took Rocío to the spot Mark had picked out, and I shared with Rocío a bit of the story of this cascading mountain stream and how it originated from springs up the mountain and then followed a course down the mountain. I then mentioned that during the dry time of year, a section of the stream dried up as the water table got lower. I told her, "Just downstream there was a hole where the water went down, and then it reemerged behind our house." She looked at me wide eyed and emphatically said, "Take me there." Mark and I exchanged a glance and with our eyes asked each other, "What's this about?"

We walked along the stream until we got to the place where there was a large hole and, since it was the dry time of year, all the water was pouring down and disappearing underground. Rocío was amazed by what she saw and proceeded to explain that this was a pugyo *(pronounced poo-yo), which is a portal to underground waterways. She revealed that these underground waterways were a means to connect and communicate across large distances and that pugyos were joined to one another. Long ago people often built shrines or temples near them, and this is one of the ancient ways for all life to be united. Then she described how when the conquistadors came to the Americas, many of the pugyos closed up to protect this primal form of transmission. She said we must do a ceremony to open our Sweetwater pugyo so that a catalyzation of communion would, once again, be available within the vast web of life. She impressed upon us that if we opened this pugyo together in ceremony, then Mark and I would become the stewards of the pugyo, which we would then tend to and feed in order to keep this portal alive.*

After the class was over, the three of us went to the pugyo to engage in our ceremony. Rocío had brought with her a small bottle of ayahuasca (her main plant ally) preparation, which we rubbed on the soft tissue on the inside of our lips. This was a homeopathic dose, which served to bring us into alignment with the profound wisdom of the omnipotence of Nature. Rocío spoke a prayer to the pugyo, and then she instructed me to cup my hands and put spring water in them as she poured a small amount of ayahuasca in my hands to mix with the water. Then I poured the water down the hole, stating my commitment to tend the pugyo and to ask for the pugyo's blessing to do so. Once I was finished, I stepped back, and Mark came forward to do the same and then Rocío. I withdrew to a rock ledge when I began to hear the most beautiful singing. I thought to myself, "How sweet it is that folks from the class have lingered and are in the field on the other side of the forest singing." I raised my head to listen more fully. Rocío saw me listening, and she said, "Oh, so you hear that." I nodded. She then said, "The spirit of the forest is pleased with us."

This experience continues to be one of my most miraculous and profound contacts with Nature consciousness where I was completely merged with All That Is. The incredible revelation that I am *not* separate from Nature is some of the deepest healing I have ever received, and I am eternally grateful to Rocío for facilitating this. No longer is nonseparation a concept for me, it has become my living reality.

Close Encounters

For years I've tried to understand what causes change in a person, and I have realized that it is one thing to change your mind and totally different when you have a change of heart. You can change your mind several times a day, but a change of heart suggests that you are growing into your true essential nature, and this kind of change is more long lasting. You have been touched on a deeply personal level, and the change becomes a part of who you are.

When one has a close encounter with Nature, usually one's heart is involved, and this is when the exchange becomes personal and the relationship changes. In this moment of synergy, separation with Nature dissolves and co-creative partnership becomes possible.

While in Belize I had traveled by catamaran to snorkel off one of the cayes (pronounced keys) that is near one of the largest barrier reefs in the world, second only to the Great Barrier Reef in Australia. I have always been afraid of big stretches of open water, so it's difficult for me to allow myself to relax while in the ocean. But this time a close encounter changed my relationship with the oceanic water world and its inhabitants.

We had docked the catamaran off of a less-frequented caye and were alone in the water to snorkel around and see the beautiful fish and coral. At one point, I found myself surrounded by thousands of little fish about two inches long who had an iridescent shimmer to them. I was mesmerized by how the sunlight reflected off them, and they seemingly weren't bothered by my presence among them. It was one of those WOW moments. I spent a long time treading water in the middle of the ensparkelation *(a word coined by my husband Mark) created by these tiny fish. Suddenly, they began to do the "one mind" move known as murmuration, where they moved completely in sync with one another, and I joined in. I found my body moving in sync with theirs in a completely relaxed state, and the ocean began to feel so familiar. I had never felt this comfortable in the sea before. Because I allowed myself to relax into the safety of this school of fish, I was able to have a close encounter that healed the lifelong separation I had felt with big water.*

Sometimes in our close encounters we experience a level of intimacy that goes beyond what we can fathom is possible in the nonhuman realms. My former student Dori's story with the Manzanita tree exemplifies this.

When I arrived at Mountain Home Ranch in April, I felt something tugging at me, like a little kid pulling my sleeve. It was calling my name the whole drive up. Once I settled in, I took a walk and followed the call and landed on a hillside of scrubby, gnarly, beautiful Manzanita. I touched her leaves, caressed her peeling, almost pink limbs like a lover I hadn't seen in ages. I whispered to her, "Are you the one who's been calling me?"

I saw myself in her—this tough survivor, fire-loving sister who also bore delicate pink bells in the spring and tiny pungent berries in the summer. Manzanita is fierce but also unafraid to be tender and vulnerable. As I sat with her, she told me her story and gave me her song. I drank in her medicine, the medicine of shedding skin and surviving trauma, only to become more oneself. We exchanged breath, and I let her in as she let me in. If you're thinking it sounds like sex, you're right. It's like being in love and never wanting to stop connecting . . . like wanting to touch every bone and cell and flower . . . like prayer and magic and being full of life.

After our first encounter, I could call her name, sing her song, and feel her healing move through my blood in the twists and turns that recall her tree body. I could feel her warmth, from basking in the sun all day and surviving wildfires, rise in the palms of my hands. I would lay my hands on the hearts and hips of my clients, giving them the deep transformative healing of Manzanita, the lady who blossoms from fire.

What I have come to know about close encounters is that they happen more genuinely if we are in gratitude and approach the present aspect of Nature with respect and honor.

My friend Becca and I went up the mountain behind my house to harvest some Ghost Pipe, a waxy white parasitic flower that is quite prolific when you reach a little higher elevation. It was one of those days where I felt pressed for time with much on my to-do list. We walked briskly up the mountain, and I began to look for the little white upside-down pipes. I saw one here and one there but certainly not enough to harvest. I found my annoyance increasing at the fact that it was taking so long to accomplish our task. Then in a flash I realized why Ghost Pipe was not revealing himself to us. We had not taken the time to ask permission to harvest and to be in gratitude for the many gifts that Ghost Pipe carries and to honor this very special plant that in many areas is considered at risk. We stopped and made offerings and prayers and said how sorry we were for forging ahead without giving due respect. Then we started up the mountain again. Within a short distance, we saw them, clusters of five to ten pipes spread over a huge area. Again we thanked Ghost Pipe for showing his face and allowing us to harvest some of him to make medicine.

Taking Action

A very crucial piece of healing the separation with Nature is to take action in order to move into nonseparation. It is important to approach our healing with gratitude, respect, and true intention.

Damanhur, an ecocommunity in the foothills of the Alps in Italy that I mentioned in chapter 1, not only includes co-creative partnership as central to its vision, the members also actively practice aspects of co-creative partnership on a regular basis. The Damanhurians made a pact with Nature where they agreed to certain conditions in their co-creative covenant. They have a dynamic connection with the nature spirits of the land, where equality and respect are acknowledged. They listen to the nature spirits of the land and then together agree on a course of action. Sometimes they may ask for help with a particular crop being grown and then honor the nature spirits for the guidance they receive.

One of the predominant aspects of the Damanhurians' pact with Nature is honoring the elements by diligently following certain criteria. For example, they respect the air by not smoking, they respect Earth by not adding chemicals to the soil, and water is honored by returning it to its original blueprint via prayers, positive thoughts, and love vibrations.

The Damanhurians have a designated area for rituals where offerings are made every week, and each week prayers are given to the spirits of the land. Their commitment is such that they consistently make a harmonic connection with Pan (the Greek god of the wild), nature spirits, elementals, and all of Nature.

I have not made an official pact with the spirits of the land at Sweetwater Sanctuary, but I certainly have an informal pact that I see as akin to marriage vows. To me a marriage vow is a commitment to your beloved to honor, cherish, and support them to be all they possibly can be. This is exactly how I feel toward Nature and am so grateful to have this level of partnership.

Much like at Damanhur, I have a particular spot where I make prayers for the spirits of the land. It is a Maple tree that I call the prayer tree. Whenever we are making prayers, rituals, or ceremonies with the tree, we adorn the tree with gifts or prayer ties. In this way I am honoring Nature and also listening to what Nature has to say about the actions that need to be taken to heal the separation with Nature.

Sacred Earth Activism

The phrase "sacred earth activism" implies that there is a spiritual aspect to activism or that the divine is being taken into account. For me this means you take action because it is the right thing to do without expecting a certain outcome. Working for Earth rights is another form of activism. However, when activism is rooted in the sacred, we automatically take up our place as stewards of Earth, and it is our *obligation* to honor the rights of Nature. When we have taken up our role as stewards, being an activist and taking actions to protect Nature are an inherent aspect of our true essential nature, and we are not "fighting": we are doing what is necessary to keep life alive.

Probably the biggest action taken to heal our separation from Nature was the creation of Earth Day. The first Earth Day was observed on April 22, 1970, and came about through the efforts of many people but most especially those of the coordinator Denis Hayes. The intention of Earth Day was to support the work of environmental protection. The first Earth Day was primarily observed in the United States, and twenty million people across the country attended tens of thousands of organized events. The first Earth Day is considered the largest single-day protest in human history. In 1990 Earth Day became an international event, and now more than a billion people across the globe engage in Earth Day actions each year. In 2009, the United Nations General Assembly adopted April 22 as Mother Earth Day, which contributed to the launch of the UN's Harmony with Nature committee.

Courtship with Nature

A powerful way to take action to heal our separation from Nature is to court what you want to be in a deeper relationship with. The art of courtship seems to have been lost in modern life, where seduction to get what you want dominates. During a courtship you courteously approach another, making an effort to see what they see, to walk in their shoes, with the intention of touching their heart with your heart. Ultimately, courtship can lead to the process of falling in love or becoming beloveds.

When you court Nature, you engage in what the Spanish call *duende*. *Language Magazine* describes it this way:

> *Duende* or *tener duende* ("having *duende*") can be loosely translated as having soul, a heightened state of emotion, expression, and heart. The artistic and especially musical term was derived from the duende, a fairy or goblin-like creature in Spanish and Latin American mythology. *El duende* is the spirit of evocation. It comes from inside as a physical/emotional response to music (or falling in love). It is what gives you chills, makes you smile or cry as a bodily reaction to an artistic performance that is particularly expressive.

Often, during courtship we serenade another to touch their heart. In my love affair with plants, when I sing or play an instrument for them, I notice they have a favorable response. When our courtship is filled with duende, we pluck the heartstrings of our beloved, creating a ballad of the soul.

Reciprocity is a form of action that is essential in healing our separation from Nature. When we give back, it is a way to say thank you for all that we have been given. Giving back is a way to feed the spirits of Nature. The act of feeding is far more than making an offering. When we feed spirit we recognize that spirit needs to eat to stay alive just like we do. So we feed spirit what spirit finds to be delicious, and what is most delicious is always beautiful. How, then, do we make beauty to be in reciprocity with spirit in Nature? We make beauty with what is unique to humans—our hands, which have opposable thumbs. Spirit *loves* when we make something beautiful with our hands, which could be food, beads, a painting, or any gift you are inspired to create. This gift is not for human consumption; it is for spirit, and ideally, you give your *most* beautiful creation to spirit because when you feed spirit then spirit feeds you. This is reciprocity. We feed each other to keep life alive.

My student Leah shares this story of how she gave back to the spirits of Nature.

> *Yesterday, in celebration of the summer solstice, I went on a Nature-inspired journey to the peak of a sacred mountain, where Celts and prehistoric communities once thrived. I brought close to my mind the independent exploration of: "Where do you most easily meet with Nature?"*
>
> *In the wilderness, where people rarely go, I sense Nature on a deeper level, especially when I'm alone, like yesterday, in moments of uneasy and unknown exploration. Sometimes, I even sense past memories of these splendid wild places. When I intently dedicate a lone journey to be with Nature, offering gifts and gratitude along the way, here we seem to connect and co-create.*
>
> *Before my picnic, at the peak, I felt compelled to sing words I've never heard in a language I didn't understand, rippling an ancient tone across the valley below. Singing is not my usual forte! I sang this song with inspired invitation and intuitive encouragement from the breezy, scaly mountain crest. Somehow, I felt I was giving back! The song filled me with such emotion, like echoes of joy and sadness were unveiling across the lands beneath. I cried, quite uncontrollably, still trying to sing the song I felt drawn to share. Images of immense forest from the distant past appeared in my mind and a sense of dragon energy! I'd never really contemplated dragons before!*
>
> *The journey was profound, ending with a three-hour steep and slippery rocky descent in thick eerie fog, which only made me ponder more. Where are the limits and boundaries between one reality or realm and another? What happened to dragons? Did they exist? Were they oppressed and suppressed by human kind? Maybe this is where my relationship with*

Nature is evolving and maturing—thinking beyond the restricted reality that my mind and societal norms create, without self-judgment.

Leah describes feeding the spirits with song. Singing doesn't engage one's hands but is definitely another way to make beauty. Along this same vein, praying is another way to feed spirit. When we pray we always speak in the first person to the spirit of the plant, the stream, the mountain, or the dawn, and we attempt to speak the most eloquent speech possible. Martín Prechtel would say, "Let your words be like honey rolling off your tongue." Prayer is always from your heart and is filled with gratitude: thank you, thank you, thank you—not gimme, gimme, gimme.

Treat Nature as if she is sacred, and she will respond to you as if you are sacred. In this way, not only does the separation from Nature heal but the separation from oneself also heals.

FOUR
Self-Love

You yourself, as much as anybody in the entire universe, deserve your love and affection.

SHARON SALZBERG

At the foundation of successful co-creative partnership is love both for oneself and other. In this chapter we are going to explore loving and nurturing self because if we don't love ourselves it is difficult to love another. When we love ourselves, we delight in sharing our gifts and shining our light, knowing that this elevates life for all, including the many aspects of Nature. As intuitive herbalist Asia Suler, author of *Mirrors in the Earth*, says, "Nature is conscious, and it is caring. It *wants* to instill a healthy sense of self-compassion within us, because when self-love is intact, we truly become better citizens of the world." We all come into this Earth walk with gifts that are unique to each of us. Some might say they don't know what their gifts are so how can they share them. One of my students, Martha, shares her guidance from Rose on self-love and sharing her gifts.

Begin with self-love. Stop all self-criticism and harshness. Allow rest. Let yourself be as you are. Love yourself freely, with an open heart. Allow the world to be as it is. Since love and Source is all around, you need only relax and open. Let it in.

Fill yourself with wonder, with the abundance, the joy of Nature. Then give. Make yourself an outpouring, holding nothing back. When you are connected to Source, you can give all, as we plants do, even our bodies and our lives—and you will be renewed, endlessly. When you are connected to Source, there can be no loss, no scarcity, no final endings. So, learn to share everything.

Know that you cannot be small in your interactions with Source. Fear, insecurity, self-doubt, anxiety, excessive concern with physical health: these are all ways of being small. So are greed, envy, grudges, negative judgments of people and circumstances. You have set an intention to become a vessel for love. Such a noble intention, my dear.

Begin by becoming tender toward yourself. Be quiet. Enter deeply into silence and stillness. Experience peace and softness. Cultivate it, like a sweet patch of earth. Nourish the softness. Feed it, as you would your own beloved baby child. This softness and tenderness is what will allow you to expand your heart as a vessel for love. Hard things cannot expand; they crack and fall apart. So begin in this place of tenderness. Remember that rest is an expression of self-love and requires trust. Learning to love the self is essential preparation for connection to Source.

Nurture Physically

The most obvious way to begin nurturing oneself is on a physical level. This may begin with changing your diet to include simple, healthy, nutrient-rich food that you grow yourself or a local farmer grows.

It is already a known fact that eating a well-balanced diet improves one's general health, increases energy, elevates mood, and heightens one's quality of life. Each person is unique and so needs to find what works best for their particular body type and environment. What nurtures me may not nurture you. When shifting to a healthier lifestyle and way of eating, pay attention to your body and what "pings" for you—meaning what gives you energy (I write more about pings in chapter 5).

The key to healthy food is healthy soil, as Dr. Zach Bush, founder of Farmer's Footprint, says, "Soil health is plant health is human health." The route to healthy soil can be found in the regenerative agriculture movement that was initiated by Robert Rodale of the Rodale Institute in the early 1980s. Since those early years of farming organically and with the health and well-being of all of Nature in mind, regenerative agriculture is proving that we don't need chemical farming to feed the world—a myth promulgated by Big Ag. The truth is we *all* can care for ourselves by eating nourishing food grown in healthy soil.

Another facet of physical nurturance is movement, which for many is experienced through exercise. For years I attempted exercise with various machines: I tried to ride, walk, jiggle, and run—while indoors—my way to a supposed healthier way of being. Finally, I came to the realization that I really hate to exercise on these machines, and I am *not* loving myself when I do this. When I looked closely at what I didn't like about exercising, I found that it felt unnatural to me. Shouldn't I just be going through my day with natural movement and walking or maybe riding my bike to get from one place to the next?

Professor Daniel Lieberman, a top evolutionary scientist at Harvard University, says, "The biggest myth is that it is normal to exercise. Our hunter-gatherer and farming ancestors never ran or walked several miles a day just for health. Still, average hunter-gatherer men and women walk about 9.5km (six miles a day), respectively, in order to hunt or collect food. We evolved to walk with extreme efficiency." However, Lieberman also points out in his book *Exercised: The Science of Physical Activity, Rest and Health* that "when people don't exercise, we label them as lazy but they are actually doing what we evolved to do—which is to avoid unnecessary physical activity." It turns out that we have evolved to rest. While some describe sitting as the new smoking, he scoffs at this notion. "Hunter-gatherers sit as much as Westerners," he says. "Studies of Hadza hunter-gatherers in Africa show them typically spending nine hours sitting, two hours squatting and an hour kneeling each day."

I am not suggesting that we become couch potatoes or that physical activity in the form of exercise is not good for us; rather, I'm proposing that we reevaluate how we move our bodies *and* how we rest so that both become a way to engage in self-love.

Nurture Emotionally

For years I have worked on shifting my internal monologue away from one of being at war with myself. Usually my internal monologue goes like this: "You aren't smart enough to understand that. You should be losing more weight. You should be kinder to Mark. Why aren't you doing more for Earth? Jeez, you are so lazy. Look at those ugly bags under your eyes!"—and on and on it goes. Basically, what I'm telling myself is not contributing to self-love and sets an emotional tone that doesn't support feelings of joy, peace, contentment, and general well-being. When the monologue starts clawing at my heart to the point at which the pain is too much, I do what always brings me back to center, I go to Nature.

I sit on the bridge that spans Sweetwater stream and watch the light dancing on the water. The undulating movement of the water and light soothes the jagged edges of my being. I feel cradled in the lullaby of light and sound from my beloved mountain stream. While in the sweet embrace of light and water, I begin to remember. I remember that I am a being of light, like the light on water, swimming through life with all its obstacles, challenges, blessings, and beauty, and I am who I am. I am me in all my shortcomings and long comings. I relax with a deep sigh as a gentle peace settles over me, and I come home to myself.

My student Martha describes her time in Nature in this passage.

I was on a mini-vision journey, on a chilly October day, at a well-loved nature spot in a park near my home. Toward the end of my time by the river, I received a surprise visit from a beautiful spirit: my own dear soul-self. She arrived, floating over me in a soft pink cloud of bliss and unity, connecting my heart to hers, and both of ours to the Great Oneness, the Holy Heart of universal love. And along with this visitation, a good old

Beatles song came too: "Bright are the stars that shine, Dark is the sky. I know this love of mine will never die. And I love her."

This visitation was accompanied by the realization that, all my life, I've searched outside myself for love. And in those relationships with humans where I've found love, I have the usual human grasping and fear of loss. Separation and death are inevitable; loss is woven into the fabric of life. It is one of our most potent teachers. The relationship I have with my soul-self is different. This is my one true love that will never die or leave me. And I love her.

Nurture Mentally

It turns out that the warring mental chatter not only causes me heartache, as described previously, but it actually can direct the expression of my DNA. Dr. Bruce H. Lipton in his book *The Biology of Belief* postulates that we are affected more by our environment and beliefs than we can even imagine. A description on the cover of his book summarizes his research.

> The implications of this research radically change our understanding of life, showing that genes and DNA do not control our biology; instead, DNA is controlled by signals from outside the cell, including the energetic messages emanating from our positive and negative thoughts. This profoundly hopeful synthesis of the latest and best research in cell biology and quantum physics has been hailed as a major breakthrough, showing that our bodies can be changed as we retrain our thinking.

In 1952, Norman Vincent Peale popularized positive thinking in his anecdotal book *The Power of Positive Thinking,* which has sold over five million copies. Peale was a Protestant minister, and much of his inspiration came from biblical sources. Now, seventy years later, science is proving what the faith-based inspirations of Peale revealed about the self-nurturance of positive thinking.

Nurture Spiritually

Spiritual nurturance can come in many ways, such as being in a meditation practice, immersing oneself in Nature, engaging in faith-based devotion, or creating beautiful art, music, or poetry, just to mention a few. Ultimately, I find that the most profound spiritual nurturance is when you walk your soul's path, meaning that you do what you came here to do during this particular incarnation. When you engage with life in this way, you are in alignment with your purpose, and all else falls into place, including any practices that support you spiritually.

We will explore one's purpose as a way to engage in self-love later in this chapter.

Good Vibes

One of the greatest scientists and inventors of the modern era who helped develop alternating current electricity that supplies most of our homes, Nikola Tesla, is believed to have said, "If you want to find the secrets of the universe, think in terms of energy, frequency, and vibration." He helped us to understand that everything carries vibrations: physical matter, thoughts, emotions, spirit, Nature, songs, water, Earth, animals, humans, plants, trees, fungi, stories, fire, stones, words, air, minerals, beauty, love—everything in life is vibration. I have given a sketch of some different vibrational resonances to tune into, but remember *everything* has a vibration, including greed, arrogance, pollution, plastic, violence—aspects of life whose vibration you might not want to surround yourself with.

Ideally, we surround ourselves with vibrations that feed and nurture us because "good vibes" go into making up who we are. As I've discussed already in chapter 2, at the nucleus of the cell (DNA level) are biophotons and biophonons (light and sound) that carry vibratory resonance. This light and sound is affected by external vibrations, as indicated by the research of Dr. David Muehsam and Dr. Carlo Ventura who wrote:

> Thus, human feelings, thoughts, psychological attributes, and perhaps even life choices may resonate with the molecular cellular level and affect even these most subtle processes of life. Although the idea that higher-level activities such as thoughts and feelings could affect gene expression may seem radical, substantial evidence exists supporting just such a confluence of psychology, neuroscience, and molecular genetics.

And from an article in *Scientific Reports* by Mobin Marvi and Majid Ghadiri: "DNA can be affected by acoustic, electromagnetic, and scalar waves. Under the influence of these waves, the genetic code can be read or rewritten."

Recently I read an article in *Scientific American* titled "The Hippies Were Right: It's All About Vibrations, Man! A New Theory of Consciousness" by Tam Hunt. Being an old hippie myself, after I finished reading through many scientific abstracts about vibratory resonance and DNA, as mentioned previously, I chuckled to myself and thought, "Well I could have told you that." And then, voilá, I saw the article in *Scientific American* and felt vindicated.

When we surround ourselves with what makes us feel good, it literally goes into making up who we are. This happens at every level: the food we eat, the clothes we wear, the music we listen to, the people we hang out with. It includes our workplace and home and outdoor environments, along with our internal environment of our thoughts, emotions, and vital essence. To me this is the ultimate in self-love, placing oneself in the presence of "good vibes" or that which plucks the harmonic chord in the symphony of one's soul.

Sovereign unto Self

Surrounding oneself with good vibrations as an avenue to self-love initiates sovereignty. Sovereignty is freedom from outside control, which is basically the freedom to be me. I came to sovereignty when I realized that I was the author of my own experience so that made me an authority. No one can be me better than me. Within my own authority is my independence, and my independence is entwined with interbeing. Interbeing does not diminish my sovereignty but instead enhances the freedom to be me since from within interbeing the brilliance of our individual gifts shine brightly. When we fully embrace our sovereignty, we can trust our luminous nature and our participatory access to Source energy. Shamanic practitioner Sandra Ingerman puts it beautifully when she says, "I honor my power to have my own direct revelation."

Direct revelation could be seen as experiencing guidance from Nature (or Source, God, All That Is, Great Spirit, Holy Heart) in various ways. What follows is a direct revelation that my student Michele had with Magnolia.

The first time I met Magnolia was in Georgia. Magnolia smiled and shared: "This is why you chose to wait and do this breathing cycle now at this very moment. You are feeling all those who have come before you and who have been oppressed: the slaves, the women, the Indigenous people. They seek comfort in me, and I offer them this spiritual healing from their torment. You connected with me spiritually in California. Can you feel now the tears of sorrow of those Indigenous people there who were displaced? Help me heal their sorrow!"

Such joy in my heart that Magnolia has chosen me. I cry grateful tears that I might be able to lessen their sorrow. I merge with Magnolia, allowing our breathing to become one. I stand up and allow my branches to sway in the breeze. I smell the aroma of my flowers and feel their sturdiness with the cones rising in the center. I feel the strength of my trunk and feel my roots firmly planted in Mother Earth's soil. I am grounded in Magnolia's strength. I smell scents of Thanksgiving spices when I connect with Magnolia. The aroma surrounds me and creates a warmth and comfort within my spirit. I am connecting with Nature as part of her through Magnolia—through Source. I find myself rolling onto my side in a fetal position. The sky opens and rain falls so hard on the rooftop of this cabin, and I hear the rain as percussion, accenting my breath. Now I am the cone inside the sepals of Magnolia, and I am growing larger and larger. There is no sense of time, almost like a time-lapse camera. The seed lands on Mother Earth and the rain waters the earth around me. With no sense of time, I am growing from seedling to a full-grown, majestic Magnolia tree. I am swaying in the breeze, and I am so overwhelmed with gratitude for this gift from Magnolia. I am crying with joy of life—everything is sacred. I am in awe of the connection we have with Nature. My breath with Magnolia's breath is one, and she has released the oppression I didn't even realize I had been feeling. Magnolia in her majesty released the sorrow that pains me for the actions of our ancestors. I am lighter, grateful, and so joy filled.

When you step into sovereignty and allow yourself to be your own authority, you stop questioning everything you receive via direct revelation. The phrase "maybe that's just my imagination" dissolves into a knowingness that your imagination is a vibrant doorway to the multiverse. Here, in this space of knowing who you are, your purpose begins to reveal itself.

Purpose

We all come into this Earth walk with inherent gifts to share with the world. When we share them, this is what makes a vibrant, purpose-filled, thriving journey called life. Sharing our gifts is also a form of self-love because we recognize that we have something to offer the world, and it is love that makes the world a better place. When I love myself enough to live according to my true essential nature, walking the path I came here for this time around, then *all* of life thrives, not just me.

For the longest time I thought I had to do it all, be everything to everybody: run a successful business on my own, do all the household tasks, be a supermom, know every single plant, be the perfect mate, fix things when they break, make sure the gardens are watered and weeded and the paths are clear—and on and on. Then I became exhausted, and I realized that I can't do it all. And by trying to do it all, I was diluting what it is that I do best and what I'm gifted at. *And* I was also keeping others from doing what they are gifted at. What a relief to focus on my inherent gifts instead of trying to be gifted at everything! Once I stopped trying to do it all, I could hone in on my gifts, which led me to my purpose.

This is one end of the spectrum. The other end is not knowing what your gifts are and thus not being clear about your purpose.

If you need help with knowing what your gifts are, ask yourself these simple questions:

- *What makes me feel good? Internally and externally?*
- *What makes me zing?*
- *What gets my juices flowing?*
- *What is in alignment with who I am?*
- *When do I feel centered?*
- *What gives me energy and, conversely, what drains my energy?*
- *What inspires me?*
- *What makes me so excited I don't want to stop or I can't wait to get back to it?*
- *How do I create and/or express beauty?*
- *What gifts emerge from this beauty?*

Write these questions in a journal and set aside a particular time of day to start answering them. Let this become a self-love practice. Be aware of how your answers may change depending on your circumstances or environment. Allow yourself to begin to weave some of your insights into your Earth walk that become a part of how you walk in the world on a regular basis. As you continue in your practice, what emerges? Do you notice a softening in your heart? Or perhaps some fluttering? Do you remember that feeling? You may find the inner monologue shifting and that small, soft voice saying, "I think I'm falling in love. I'm falling in love with myself."

Once your purpose comes to light, you realize it has been here with you all along. However, actually living your purpose comes with the caveat that you have to make choices along your path to do so. The Cherokee story of the two wolves could be applied here when it comes to making choices.

An old Cherokee is teaching his grandson about life. "A fight is going on inside me," he said to the boy. "It is a terrible fight, and it is between two wolves. One is evil—he is anger, envy, sorrow, regret, greed, arrogance, self-pity, guilt, resentment, inferiority, lies, false pride, superiority, and ego."

He continued, "The other is good—he is joy, peace, love, hope, serenity, humility, kindness, benevolence, empathy, generosity, truth, compassion, and faith. The same fight is going on inside you—and inside every other person, too."

The grandson thought about it for a minute and then asked his grandfather, "Which wolf will win?"

The grandfather simply replied, "The one you feed."

This story depicts the inner conflict that exists within all of us, but primarily it is about which path you choose and offers insight into how we can live our purpose. Do you choose the path that feeds your purpose, or do you choose one that diverts you from your purpose? This is part of the human condition—we must *choose* to live according to our true essential nature.

Love Originates with Beauty

Dr. Zach Bush, one of the great inspirational leaders of our times, says, "We are designed to express beauty," and he suggests that love is not the most powerful force in the universe but that it is beauty because this is where love originates. I have contemplated this for quite some time as I am a person who attempts to surround myself with beauty as much as possible. It is not only beauty in my outer world but also inner beauty, which is more associated with a quality than an object.

Beauty is everywhere if we only take the time to see, feel, and absorb it. How do you feel when you are immersed in beauty? Physically, I feel warmth in my heart; emotionally, I feel joy; mentally, I think positive thoughts; and spiritually, I experience connection.

Standing at the water's edge I am bathed in the beauty of the full moon rising up from the sea. A path of silver light spreads across the expanse, finding its way straight to me. My breath catches as I gasp in awe of this river of light dancing in rhythm with the great life all around. Can I hold such beauty, is my container big enough, am I worthy of such a gift of sheer luminescent joy? I find myself starting to sing a Brooke Medicine Eagle song from her Visions Speaking *album that has lived inside me for decades, "I walk a path of beauty, walk a path my ancestors laid out before me. I walk a path of beauty, walk a path my ancestors laid out before me. Oh, yes, I make a path of beauty, hold my visions, lay my dreaming out before me. Oh, yes, I make a path of beauty, hold my visions, roll my dreaming out before me."*

The light flashes, and I realize I have been walking this path my whole life and it has led me to this Now moment. Here I am in all the exquisite beauty of light on water, with the path of light leading to me. Here I am, and I am beautiful.

In my years of engaging in co-creative partnership, I find myself encouraging my students to work with aspects of Nature that, seemingly, are not so easy to partner with. It's one thing to partner with a beautiful flower or bubbling mountain stream or majestic Redwood tree, but what of plants that are poisonous or give you a rash, of people who are abusive, or environments that have been polluted? How do we find the beauty or gifts in these encounters? In my own journey I have found that having the commitment to look for the beauty and gifts in every aspect of life is of primary importance. From here, I focus my attention and intention on discovering the gifts and allowing the beauty to flow while setting judgment aside.

As Bush so eloquently expresses, "The beauty around you is dumbfounding and if you would just be witness to it for a moment you will experience the most pure form of love you ever have. In that, you will find a reverence for yourself and a reverence for life." He suggests that because beauty drives a greater expression of life, when we fail to see and experience beauty, life will end.

As we step into self-love, we see how this act inspires love of Nature and how love becomes a verb instead of a noun. We begin to experience being love, which is very different from being in love. In this state of being love, the elegance of all life is unmistakable. Here in the miracle of self-love, we can be all we possibly can be, in all our beauty, living in harmony with Nature.

FIVE
Trust

To not trust the touches from Nature is like disrespecting my grandmother.

Pam Montgomery

Trust is a key ingredient in any good partnership because it engenders honesty, sincerity, and lack of harm. When you do not doubt the aspect of Nature that you are in partnership with and you listen to the guidance being given without judgment, this serves as a huge affirmation, validating both oneself and Nature as co-creators. It is important to make the distinction between trusting in Nature and trusting in people. Nature is living according to its own true essence, not putting on different façades depending on the situation. Some might say that camouflage found in Nature is a form of putting on a different façade, but I maintain that this is part of the true essential nature of the frog or octopus or chameleon or stone plant (*Corydalis hemidicentra*) and is an adaptation for protection from predators.

First Voice

The big question that comes up around trust when one is in communication with an aspect of Nature is: How do I know the guidance is coming from the plant? Isn't it me making it up? The real issue here is not whether you can trust Nature; it is whether you can trust yourself. In my own process of trusting the guidance I receive from Nature, I rely on my first voice. First voice is the very first words, feeling, or sensation that comes when I'm communicating with Nature. Your first voice is your inherent knowing, that which is intrinsic to you. First voice is something you naturally and instinctively know. Sometimes this is referred to as primal knowing, which is original and, again, comes first. Inherent or primal knowing is your portal to the vast interconnected web of biointelligence.

Many feel that they need to train themselves to hear their first voice because they constantly second-guess themselves. I feel like one's second guessing comes from one's ego. When these moments arise, my teacher Martín Prechtel used to tell us "to send our egos out for donuts." I experience my ego as "the nag" who sits on my shoulder and literally nags me about this, that, and the other. I ask the nag to go sit in the corner until I'm done with my exchange with Nature. Sometimes the nag needs a distraction, so I ask the nag (my ego) to take notes on my communication or experience or daydream. This is a perfect job for the ego, since note-taking and list making are gifts intrinsic to the ego. I tell my ego that I'm going to listen to my heart for my first voice—the very first hit that I receive—and that I want her to record that first voice of primal knowing. In this way, my ego feels helpful and not as if it is being disregarded or dismissed.

Some may refer to primal or inherent knowing as intuition, or that which is known without conscious reasoning. Intuition is a sense that we *all* have; it is not just for the chosen few. The key to developing intuition is listening. If you are a good listener then you are likely adept at paying attention to your intuition. One of the things I suggest to my students is to listen with their "big ears," which means to listen with all five of their senses. What I have found over the years is that by deeply attending or paying attention to *everything* that is going on in and around me, I can read the energy. When I do this, it jump-starts my intuition, which I then mingle with the morphic resonance or collectively held memory of the Nature ally that I'm communing with.

What follows is an excerpt from the foreword that I wrote for Carole Guyett's book *Sacred Plant Initiations.* It has to do with a story I touched on earlier, in chapter 2. There are many aspects of co-creative partnership in this excerpt but I am including it here to illustrate trusting, tapping into one's heart, and being in the flow regardless of external events. We will go into depth about plant initiations later in part four.

> My first plant initiation began in the spring of 2010 while I was in rural England on a work assignment teaching Greenbreath and plant communication to a group of interested students. The plan was for a relatively simple and consolidated round trip, including time on the front end to touchdown in Ireland for a few precious days with my so distant yet close friend Carole Guyett. Days later in the course of teaching in England, I happened early one morning to take notice of a vast array of little yellow Primroses painting the hillside behind my lovely little guest cottage in the Cornwall countryside. I had never seen so many in one place, certainly not at home in Vermont where *Primulas vulgaris* ordinarily lives as a cultivated plant instead of one wildly covering hillsides.
>
> The time passed well. However, when winding up the class and departing for London for my return the next day to the States, all civilized plans began to turn to dust (or ash as the case might be) for the news on the wind was that a powerful volcanic eruption was occurring in Iceland and spewing tremendous ash clouds into the air. Due to shifting air currents over land and sea, airlines were becoming powerless to fly as scheduled. On any given day whole airports were shut down so that European and international travel came to a halt.

I had only one night booked in London and once I found that I would not be flying home the next morning, I went into a mild panic wondering where I would stay. Thousands of people were in the same predicament trying to go somewhere but were unable to. After many hours of tearful communications back home with my husband, Mark, we decided the best plan was for me to try to make my way to the coast for a boat back to Ireland where Carole and her husband Steve might provide much needed assistance and comfort for me under these circumstances. I was scared, exhausted, and worrying about getting home for my apprenticeship course beginning five days later. If I could just get back to Ireland, I would at least be comfortable even if I was stranded in a foreign land.

All the boats sailing across the Irish Sea were booked except one middle-of-the-night boat that traveled to the southern port of Rosslare. It was the only possible way to get back to Ireland and so I decided I would try to make it. But I was in London, a country girl in a city of eight million people trying to negotiate getting to a boat to take me to Ireland with no cell phone, map, or helping hand. I inquired at the hotel desk how to get to the train that would take me to the boat that would take me to Ireland. She calmly pointed out the door and said, "Stand on that corner and wait for the red bus." Do you know how many red buses there are in London going to innumerable different locations? As I stood there watching one red bus after another go by with my two suitcases, backpack, and large frame drum hanging off my shoulder, a large tear found its way down my cheek. I knew it was time to make a shift, and as my friend Rocío would say, to begin to "manage the forces." I thought of all those little Primroses on that English hillside and a smile came to my face, a deep breath followed, and I gathered myself to ask for help.

My initiation had begun.

I stood at the bus stop and called on my "tower of strength" ally, White Pine. "I need help here; I really can't do this alone." Suddenly an angel appeared in the form of an English businessman of Indian ancestry. He asked me where I was going and I said Ireland. He proceeded to tell me this bus, that train, then another train, get off here, go there, and then find the ferry. I looked at him with what must have been a shocked wide-eyed stare, at which point he said, "I'll take you to the train." As I stood, completely at his mercy, I heard the words of my mother, "Don't ever go off with a stranger." I gave him a quick scan and checked in with my heart—it went "ping." I got on the bus with him and off we went. He safely delivered me to the train after the bus ride and helped me buy the correct ticket to the boat that would deliver me to Ireland. He walked to the platform with me to make sure I was on the right train and then said, "Goodbye and good luck." "Wait," I said, "please, what is your name?" "Hans," he replied, and I said, "Thank you, and my name is Pam." He turned and was gone. I got on the train and sat in a daze wondering who he *really* was, a British-accented Indian named Hans—wow!

After three trains and an all-night ferry across the Irish Sea I arrived in Ireland at 6:15 a.m. to find Carole's husband, Steve, waiting for me on the dock (after a three-hour drive from his home in County Clare), saying, "Can I take your bags, mate?" I fell into his arms in a heap with the most gratitude I had experienced in a long time.

Once back at Carole's home she immersed me in preparations for a Primrose initiation she was offering for Bealtaine (or Beltane) on May 1. We roamed the lanes and countryside picking beautiful little flowers with heart shaped petals to make into an infusion for the plant elixir. I had the great fortune of helping Carole every step of the way and when it came time to make the elixir, I also was privy to her very conscious intentions to prepare the dieting mixture in a ceremonial way. I felt I was being given a great gift by being at the right hand of a high priestess. After making the preparation she gave me a bottle of it to bring home so I could have my own ritual at the same time she and her group would be doing their diet in Ireland.

So much happened during my solitary Primrose diet! I knew I needed to rededicate my life to the plants, deeper than I had ever gone before. It was time to fearlessly walk into the unknown, trusting that my power is my love, my love for the Earth and all her beings, especially the plants and the trees. Just as I began my initiation in an unknown circumstance with an unknown person in an unknown place and an unknown timing for when I would be home, I trusted my plant allies to help me and they did. Primrose is known as the key flower and she surely gave me the key to my greatest strength.

When I look back on this experience, I realized that if there hadn't been a volcanic eruption while I happened to be in England, I would never have had the opportunity to be with Carole while she prepared for the Primrose initiation and I wouldn't have done the initiation myself. I stepped into the dreamtime of Primrose before the volcano erupted, and as I wrote in my journal, 'I'm walking in a spiral path with a clear quartz stone in the center. The quartz is like a beaming station beaming me, with the help of Primrose, to my White Pine tree at home. My energy body is preparing to travel home, physically home, but more so, home to myself." Later I wrote, "these plants are asking us to investigate the questions, who am I and what am I doing here? They are pushing us to be who we truly are, to be truly human." This was a profoundly deep experience of a plant urging me to step fully into my inherent creative power of being human, and realize that, as Carole says, "The plants can help us to develop as mature human beings and truly walk this earth as sacred humans."

Heart Ping

Several years ago my dear friend Kate taught me the ping/thud method, which is a way of working with one's heart to determine answers of yes or no. This method is quite similar to the positive/negative response technique used in kinesiology. When using the ping/thud method one needs to pay close attention to the physical sensations in one's heart. In the example in the story above, I did not have time to contemplate what my heart was saying. I went with the very first thing that came, which was a ping that gave me energy. I felt the big *yes*, and my whole being was flooded with positive energy. If the response had been no, then I would have felt a thud or that my energy was being drained. I might have experienced an actual change in posture where my shoulders drooped. When you first begin to work with this method, you may need to take time to practice identifying the felt sensations of ping and of thud, but once you have it down, this method is an effective and quick way to receive answers in the moment that don't require a big, long explanation. Of course, the key is to *trust* the answers you receive.

Heart coherence, according to the HeartMath Institute, is a state of cooperative alignment between the heart, mind, emotions, and physical systems that expands into personal, social, and global coherence. I mention heart coherence because being in a state of heart coherence can lead to trusting in your heart when you receive pings or thuds and trusting yourself, others, and Nature.

Alleviant Integrated Mental Health is a mental health organization based in Arkansas that focuses on an integrated, whole person approach to mental health centered upon the brain and heart coherence model. Alleviant has identified on their website twelve benefits of this coherence state:

1. Feeling Whole
2. Feeling Unity with Yourself and the World
3. Deep Inner Peace
4. Decreased Stress
5. Increased Energy & Vitality
6. Enhanced Creativity
7. Emotional Intelligence and Mastery
8. Greater Resilience
9. Solid Mental Health
10. Improved Physical Health
11. Deeper Emotional Connection with Yourself and Others
12. Improved Quality of Life Overall

The Yes Vibration

Our trust in what is being shared with us by the current aspect of Nature we are working with is a huge affirmation that gives validity to our partnership. When we say yes to interactions with Nature and live in a state of conscious awareness that Nature carries vast intelligence, which we have access to, we can relax and let go of our "crown of creation" syndrome, a long-held belief that humans are vastly more intelligent than all other beings. There are sentient beings within Nature, not the least of which are plants, trees and fungi, that are worthy of our respect, partnership and love. When we say yes to them, our union with the natural world is like a homecoming, and we are never alone ever again. Experiencing the depth of Nature's love and connection for us is like donning a cloak of exquisite beauty that wraps us in a creative potential to be all we possibly can be. Imagine that!

Breathe deeply into your heart. Feel the yes vibration flooding into your heart. Feel the tingle and luminescence of yes. No obstacles, only the yes-stream. As you go with this flow, feel yourself relax as you step out of the mainstream and into the yes-stream. Feel the difference between these two streams. Picture yourself in the yes-stream as all of your inherent gifts are revealed to you. Marvel in the miracle of you. See yourself sharing your gifts with people, animals, plants, water, land, all of Nature. Feel the precious brilliance of the yes-stream and how the longing to be in this state has been with you for many soul-returns to Earth. Here you are, now, living according to your original blueprint, living according to your true essential nature. Know that you can trust yourself and Nature to be microscopically honest with each other because there is nothing to hide. In this potent force field of unity consciousness, there are infinite probable realities that are life-giving. In this vast yes-stream of consciousness the love vibration is predominant. Feel the deliciousness of being love. Take one more deep breath and allow yourself to marinate in this space of the yes-stream.

Being in the yes vibration doesn't mean that you can't set boundaries when necessary. You can do this within the yes vibration simply by stating, "I say yes to setting healthy boundaries for myself." The vibrational message is one of doing something positive for yourself, which is very different from saying no, which sets extensive limitations on your actions and attitudes.

The yes vibration is a highly magnetic frequency. If your predominant energy is that of yes, then this is what will come back to you in your life. This is the Law of Attraction that we saw in chapter 1 in my discussion of the Twelve Laws of the Universe. Gary Quinn in his book *The Yes Frequency* says:

> I had a client who was in a bad situation. I told her to repeat YES a thousand times every day for seven days. Guess what happened on the seventh day? Miracles started happening—energy shifted, everything changed, because she altered her subconscious mind. She shifted her frequency, and by doing so was able to receive positive changes. After those seven days, she said to me, "If I can do this, anyone can."

Sacred Trust

When we speak of trust being sacred, we enter the realm of the holy where we dedicate our life to that which we consider to dwell within the divine. There are many terms used to reference this ineffable presence in our lives. Many people have a sacred trust with God or Goddess. Some refer to the Great Spirit with whom they have a divine association, and others may refer to the Holy Heart or Source, which they place their sacred trust in. For me, Nature is the epitome of that which is sacred, and all of the previously mentioned aspects are contained within Nature. The life-giving, omniscient matrix of Nature began in the oceans billions of years ago and continues to evolve into a more complex, intelligent, and beautiful consciousness that I refer to as All That Is.

In my covenant with All That Is, I trust Nature with all my heart to be in a reciprocal relationship of sacredness with me. This may sound like fancy semantics, but my everyday reality is that I reside within a living testament of the expansive and generous spirit of Nature.

Years ago I asked myself the question, "What if Nature was my spiritual teacher?" This is when I began in earnest to enter into a sacred trust with Nature. I stopped questioning if the guidance I was receiving from Nature was real, and I committed myself to waking up and remembering that I am not separate from All That Is.

Early Morning in My Sanctuary Garden

The early morning sun filters through the mist in a prism of light that embraces the wet warmth of budding spring. The diamond-like particles of life force are so tangible that I can sense the rhythm of my heart beginning to ride the waves of harmonic syncopation, tuning my resonance to that of the enveloping natural forces. I sit quietly allowing the touch of Nature to penetrate each pore, each sense, and each invisible thread as my cells receive the sweet caress. A warm flush of gratitude courses through my veins as the peace of the moment bathes me in well-being. In this moment of peace, the life-giving force surrounds me, fills me, and connects me to All That Is. Sitting here I know that this is my church, the temple where I can hear the sublime language of spirit speaking through Nature. I am supported fully to be filled with my own true essential nature. The deep care of the sustainer nourishes my soul so profoundly that I feel the healing taking place (see color plate 4).

Nature Sanctuaries

Another aspect of sacred trust that I have incorporated into my co-creative partnership with Nature is that of sanctuary. Traditionally, the term *sanctuary* refers to a sacred or holy place. A sanctuary is also a physical place of refuge or safety. Also, one can seek, find, or take sanctuary in another being, place, or thing. When I say the word *sanctuary* I notice that it initiates a calming effect, and I begin to reframe how I interact with land. I breathe a sigh of relief as if everything will be all right. Earth will heal, the water will be drinkable, the air will be fit to breathe, the fish will thrive, the polar bears will have ice, and the Redwoods will survive. If by some miracle we actually started thinking, feeling, and being with the land as a sanctuary, where we easily bond with Nature, maybe, just maybe, we will remember that we too are a part of Nature and that when we harm Nature we sabotage ourselves. When we relax into the nurture of Nature, our hearts open, resistance melts away, and we begin to connect to ourselves, which then connects us to the larger Nature.

A sanctuary, and all that goes into creating and sustaining it as such, serves as a mirror for us to look deeply into our heart, soul, and spirit to see where we are out of alignment, both within and without. Anne Romance, of Green Heron Sanctuary, says of her students who came to work seasonally in her sanctuary and Medicine Wheel, "Each time they came, more was revealed to them about the negative impact of the disconnection from Nature. They saw this as the single cause of heartbreak, soul loss, and spiritual malnourishment, all of which creates disease, separation, aggression, and greed. They saw that the biggest work they could do was to be in a reciprocal relationship with Nature." Within intentional sanctuaries, the devastation of disconnection from Nature is healed.

A sanctuary is a place of refuge, protection, or safety where all are free from harm. It is a haven for the land, and all who dwell and pass through there, including plants, trees, wildlife, and elementals, as well as the people who interact with it. Nature sanctuaries are governed by an anthropogaic approach. *Anthropogaic* (as opposed to *anthropocentric* or human-centered) is an emergent term that combines *anthropo*, referring to humans, and *gaic*, referring to Gaia, and means that people are engaged with Nature as a living being with equal rights to thrive. In human communities it has been statistically shown that there is a direct relationship between increased inequality and increased levels of violence. This trend seems to correspond to humans' relationship to Nature as well. The more pervasive the attitude that Nature is a commodity, the more violence is perpetrated against her. When Nature is kept safe and treated with the same equal rights that are given to humans, she responds in kind. Nature's original vibratory resonance is one of being protected and cared for while having equal rights to life. When these conditions are in place, Nature can exist in an easeful and relaxed state of well-being and harmony, not in a state of stress. When we realize that we, too, are a part of Nature, the phrase "do unto others as you would do unto yourself" takes on new meaning. Susan Clearwater of Green Turtle Botanical Sanctuary says of her students when they were working in her sanctuary, "They started to understand the interconnectedness of it all and how this brought safety and security. It empowered them to see how Nature works and how much is given to them from the plants, the land, and the elements."

Many nature sanctuaries are established for the explicit purpose of protecting the plants and trees or a particular tract of land. There may be plants already living in an area that needs protecting or plants may be cultivated to restore an area to a natural distribution of native plants. One such organization that is focused on protecting North American native medicinal plants is United Plant Savers (UpS). UpS has created a Botanical Sanctuary Network where individuals and organizations designate their land as sanctuaries that protect native medicinal plants as well as provide educational opportunities around the importance of these species and the threats they face. These sanctuaries, through the safety they are restoring to the plants, have an effect upon the people who are their caretakers. When caretakers become stewards of land, they enter into a different, more ancient relationship where the land isn't shackled and enslaved but is free to vibrantly thrive according to its nature without threat of harm.

In the words of founder Rosemary Gladstar, in an article from her website:

> By becoming stewards of the Earth and creating living sanctuaries, we build repositories of well-being not only for humans and plants,

> but for all life forms. And that helps to change the world . . . plant by plant, tree by tree, bird by bird, sanctuary by sanctuary. . . . What makes a botanical sanctuary different than any garden or backyard space? *Sanctuary* is defined as a place of refuge, of protection and safety, and a place where one can feel at peace. It is also defined as holy or sacred space. A botanical sanctuary is created as a haven and refuge for plants, native species in particular. Intention and purpose is what makes it different than any other backyard space; a botanical sanctuary is created with the intention of providing a welcoming refuge and haven for plants to thrive. And wherever plants thrive, all other life forms thrive, including humans. . . . a botanical sanctuary has little to do with size or ownership, and much more to do with relationship and intention. It is about good stewardship practices, a relationship that takes into account the natural resources of the land and its native habitats and inhabitants; it is about restoring the sacred relationship between people and land.

A sanctuary as a sacred, consecrated space in a natural setting has been recognized for thousands of years. Many of these historical sanctuaries were gardens, wild forested areas, mountaintops, or sacred groves. The trees in these sanctuaries were considered sacred because the spirits of living gods and goddesses dwelt within them. Each traditional culture who tended a sacred grove, from the British Isles to Greece, India, Africa, and Malaysia, gave particular spiritual importance to these groves and performed ceremonies and rituals there, much like a modern-day church service. During the Crusades' there were efforts to eradicate Earth-based forms of worship; it was common to cut down the groves and build churches on top of the site.

When an area is designated as a sanctuary, there is an acknowledgment that a force greater than ourselves is present and we can have direct access to this life force. Former student and biblical scholar Martha Hamilton suggests that throughout the Bible it is indicated that "sanctuary, ultimately, is where God's presence dwells," and when one is exposed to this presence they are touched in undeniable ways by the living spirit.

Another aspect of sanctuary is that of nurturance and healing. One of the primary tenets of any healing work is do no harm. Within a sanctuary setting there is clear intention that no harm will come to the plants, the land, the animals, and the people. When this principle is observed, all who are a part of or enter the sanctuary receive a deep sense of being nurtured. The level of care present engenders a feel-good experience, which then sets into motion healing on all levels—physically, emotionally, mentally, and spiritually. This healing is reciprocal: not only do people heal but the land, plants, trees, animals, and elements heal as well. There is a two-way exchange of the resonance of well-being. Tammi Hartung of Desert Canyon Farm Sanctuary tells the story of her sanctuary.

> *Twenty-nine years ago when we arrived on this piece of ground, it was very sad, indeed. There were a lot of weeds and barren earth; everything seemed thirsty and beaten down. We could feel the desperation of this land that had been neglected, abused, and disrespected in so many ways. It was screaming to us for some gentle, but serious, tender care. We had a sense of how it would respond with some nurturing. The land sent out a promise of partnership as we made a conscious choice in our sustainable lifestyle to honor that promise and become an example to the world around us. And so we began our work together to heal each other, to be a sacred place for all that live here—human and nonhuman—and all the people that visit here. This land has become a place of huge abundance to the plant nation, the wildlife nations, the soil nation, and the water nation. It is a true sanctuary that wraps our community in loving arms and offers reassuring peacefulness to all. We receive the nurturance of calm and quiet during a time that is so important to receive this gift. The healing gifts of this sanctuary seem to reach beyond its borders like tentacles to entice others to come and be nurtured also. It is amazing and wonderful to be a part of this interconnection with nature.*

Many sanctuaries have gardens as a focus because plants bring about much healing. Much of my life's work has been working with the healing aspects of plants, both physically and spiritually, so plants are a major aspect of my sanctuary space. What I have observed is that when one spends time in an intentional sanctuary space with plants that are treated with reverence and respect and are in a reciprocity relationship, they respond by offering healing to one's heart, soul, and spirit. The gardens within a sanctuary are a co-creation between humans and the plants, where the plants have an equal responsibility in the creation of the sanctuary. Alongside plants, I incorporate altars in my sanctuary garden to honor each of the cardinal directions of east, south, west, and north (see color plate 5). I place statues, crystals, or globes in each of the four directions to enhance their intrinsic qualities, which then helps to bring balance to my energy body when I sit in that direction.

As I tend my sanctuary, I engage in a living expression of my love and devotion to All That Is, which honors my commitment to the sacred. The sublime state of being in a sanctuary engenders communion with spirit, which carries the vital principle held to give life. Here in a sanctuary, our separation from Nature is healed, and I come home to myself and my Nature kin, bonding in profound ways. Here in a sanctuary the miraculous is possible.

SIX
Time

The way we spend our time defines who we are.

Jonathan Estrin

The single most predominant complaint I hear when I survey folks about what keeps them from engaging with Nature is their lack of time. It is as if time has become the oppressor that rules one's life. This kind of time, or what I refer to as external time, exists outside us and is imposed upon us by the constructs of society. External time is so essential to our present culture that it is even embedded in our language through the different words we use to refer to the past, present, and future.

External Time

The measuring of time began in the Egyptian era approximately five thousand years ago, so external time is not necessarily a modern-day invention (though I would suggest modern society has taken it to the extreme). The tyranny of external time is particularly apparent in parts of the world that are production oriented, such as in the developed world, mostly Western countries, or where the acquisition of wealth or furthering progress is seen as desirable. Benjamin Franklin named a deepening cultural trend when he wrote in the 1750s, "Time is money." Throughout the Industrial Revolution, external time gained greater rule over the business of our lives, not only in the sense of what we do to earn money but also in how we arrange our daily activities. In her groundbreaking book *Thus Spoke the Plant*, scientist Monica Gagliano writes about how concepts of time evolved.

> It was the 19th century globalizing world that succeeded at inventing and imposing a worldwide system of global time governance by supplanting local time; it turned time "out there" into a business that thrives on the collective preoccupation with ideologies of time saving, punctuality, and efficiency—the business of regulating and stabilizing the most precious of all commodities—our experience of life (time) itself.

When we live with the feeling that there is never enough time, our freedom to do what we want is reduced because we must go to work, cook dinner, take the kids to soccer practice, clean the house, run errands, or pay the bills. Our minds are constantly juggling how to take care of all of our responsibilities and fit in time to take care of ourselves: "But I need to stay healthy so I need to go for a walk, garden, get a massage . . . but can I really take time for a massage?" In our overly scheduled lives, spontaneity is all but extinct. A lack of spontaneity limits the possibility for magic moments to arise unbidden. The never-enough-time mentality creates a low level of constant anxiety that becomes background noise to the point at which we think it is normal. This stress insidiously erodes our vitality—physically, emotionally, mentally, and spiritually—to devastating effect.

The cavernous abyss we fall into when ruled by external time is one of the most detrimental casualties of our separation from Nature. As Gagliano so eloquently states:

> While promising to make many aspects of dealing with a globalized world easier, time "out there" had managed to abolish (for the

> most part) our multidimensional potentialities and possibilities by stealing time "in here." Regimented like a military action plan, the invention of time as 'out there' did well at moving humanity to a false point of departure and increasingly dislocating us away from the connection to the Earth and the natural solar cycle, thereby inducing a loss of sight of the real eventual dimension of life itself. The invention of time "out there" is the primary cause for the profound ecological crisis the planet is currently experiencing. It is a form of violence, in which time is purely human and existence itself is reduced to an exclusively human affair.

Like the rings of a tree, external time has embedded itself into the very core of our lives, and we have come up with a myriad of phrases to express the many aspects of time. Time has become a commodity: we buy time, spend time, find things time consuming, or become a big-time spender. Time is militaristic: we use the phrases killing time, marking time, or time bomb. Time runs our life, and we never have enough of it because time's a wastin': it's crunch time, time flies; we are out of time, pressed for time, or in a race against time. Time is uncertain: we are on borrowed time. We indicate the past with phrases such as before one's time, old-timer, time warp, and once upon a time. We indicate the present through phrases such as the time has come, lose no time, it's high time, it's about time, and for the time being. We indicate the future with phrases such as a matter of time, all in good time, I'll catch you some other time, when the time is ripe, or withstand the test of time. We communicate problems using phrases such as wouldn't give one the time of day, have a time of it, do time, is this a bad time?, a devil of a time—just to mention a few ways of referring to external time.

Once again, I'm grateful for Gagliano's impressive interpretation of time's relationship to the massive separation from Nature we are currently experiencing.

> The invention of time as human property that we own and control is probably one of the most sophisticated acts of planetary hege-

mony we have ever conceived, and, of course, it is a clear sign of the fundamental predicament—the deep delusion of separation—we have caught ourselves in. By replacing true time with an artificial, mechanical time that does not even exist, this device has succeeded in controlling humanity itself by abolishing (for the most part) its multidimensional potentialities and possibilities. If it is true that by controlling time you control everything, then reclaiming time "in here" is possibly the most powerful and revolutionary act of empowerment we have to bring ourselves and the whole planet away from the brink into oblivion.

Internal Time

People who live close to Earth and are still in sync with Nature operate on internal time, or what Gagliano calls "time 'in here.'" Internal time is not imposed but instead meanders like a rivulet through the nooks and crannies of one's life. Imagine a kind of time that is not a slave master but instead a trusted companion who tends to your needs and necessities but makes room for your dreams and fantasies and who is never too much but always enough. Internal time allows for our own rhythms to naturally unfold, and we live according to our own true essential nature. We determine when to sleep, when to eat, when to relax, and when to play according to our own needs. When we are freed from an externally imposed schedule, our tendency is to calibrate our body's rhythms with the rhythms of Nature so that we rise with the sun and sleep when the dark comes, which, in summer, can be as late as 10 p.m. My most favorite days are those when I have no agenda and wander through my gardens moving to where I'm attracted to and doing or not doing whatever moves me.

It is a glorious summer day, the sun is shining brightly and a light little breeze lifts my hair as I wander ever so tenderly, barefoot, through my garden. The lofty aroma of Tulsi floats on a zephyr as I am transported to mingle with the divine. What a delight Sacred Basil brings to this sanctuary! Ah, and my precious Calendula, you are so cheerful today with your radiant petals reflecting the sun's brilliance. If it delights you, I will pick some of your prime blossoms to be added to oil to make into a fine healing salve. I hear the song of Sweetwater stream and make my way to her banks where I sit on the bridge and dangle my feet over the side, drifting into a mesmerizing trance of light and sound on water. Sing me your song, sweet water, lifeblood of this land. What ripple rap revelation will you share today? I spread my body out on Earth and let the gravity of love hold me close as the soft blanket of grass and wildflowers cradle me while I deeply rest. Here, in this sublime space, I feel like there is no time, but perhaps I have entered into sacred time where spirit orchestrates the flow.

This migration away from internal time (true time) to external time (artificial time) has been one of the major contributors to separation from Nature. We have been deprived of internal time, eliminating our ability to live according to our true essential nature, which being in a co-creative partnership with Nature not only supports but catalyzes. When we sink into internal time, we easily shift into symbiosis with Nature where co-creation becomes part of who we are. Here, in internal time, we are empowered to create our own reality based on our own timing, our own sovereignty, and our own truth. When we collectively engage with internal time, we step into the vast potential that together we *all* have access to.

Internal time is characterized by spiral movement and multidimensionality, whereas external time is one dimensional and moves in a linear progression, either backward or forward. When we dwell in linear, external time, we can exist only in the past or the future, and the tendency is for the future to become abstract, existing only theoretically or as an idea. Conversely, when we reside within internal time, we have access to the multidimensional multiverse where the future weaves in and out of the fabric of time.

One of my students, Leah Black, shares her experience of internal time. This story was originally printed in the June 2023 newsletter of the Organization of Nature Evolutionaries (O.N.E.), currently available online.

Existence in Nature's Rhythm

Birds pray to a rising sun in a respectful slip of unfamiliar silence, as the village fades away with each step towards swelling mountain peaks. Four days alone in the wilds embark. No watch, phone or company by my side. Solo, free, timeless solitude. A quest for vision and meaning. A walk without destination. It is the build up to Samhain, warm air and cozy glows of crispy color. As darkness falls beings cry their eerie tales from the dimmed forest below, and the eve fire hypnotically dances to the sporadic sound of their howling and screeches. My mind rewilds with every unexpected crack and chill. Sky lit by scattered stars, uprooting a scene that takes me back to another time. Rather than sleep, I rise like a weary warrior for my evening walk. Soon drifting along winding mountain paths into the pitch black of night. Here, an uncanny character joins my silent mission. A white wiry half-wild horse. Majestic, old and full of wisdom he walks tall by my side. I bow to him like a royal soldier. He looks straight on, focused with intent towards sharp mountain edges. We march together, energy in motion.

Senses sharper, sight pierced and my walk more powerful. With the white horse now long gone, I sit myself heavily on an ancient tomb; icy quartz touching my palm and a navy night opening my wonderings. What's hidden in the earth below me? Who stood here in this sacred place to sing or cry when time existed only in seasons, skies and senses? Time, like my imagination, grows untamed as I enter weighted thoughts of ancient lives. Are these memories of another? Who thought to steal the circle of the sun, wrist-bound and wall-hung?

I feel my soul awaken and my heart open like dandelion petals spreading their golden rays at first light. Blackbirds chime the arrival of dawn as hefty clouds advance over mountain peaks, like an army of Pegasus stampeding a purple tinted sky. The light of another day, another era, fills me with awe and stillness. My journey into and out of time continues and years of anxiety begin to drop away like the falling leaves of autumn, showering their magic ochre fairy dust around me as I stroll.

Weatherscape, landscape and skyscape merge with the soulscape of my being. How hours can pass without realizing; simply staring at clouds and watching drowsy grass sway in a breeze! I am captivated, timelessly, by the horizon. The distant kaleidoscope of light, deep with natural beauty and enchantment charms me with co-created words of love that I scribe into pages of a blank book. Soon the sundial of Nature's shadows will fade again until tomorrow. Fear, once held, now withers away with the light of a second wild dusk descending.

Day turns into night, night into day and something deep within me seeks to feel thorns of a fruit bush scratch my freckled skin, in a quest of autumnal challenge and reward. With sweet berry stains on my fingers, I sit in a painted meadow weaving my prayer stick while I watch the world go by. Nothing more, nothing less. I feel "human." Fog appears like a white dragon snaking through mountain valleys, full of vigor and personality. Life-filled misty liquid dances with premeditated movement, as graceful as a theater phantom flittering through mountain cracks. I could almost jump into the fleecy ocean below me to swim weightlessly in a pool of imagination. As I stare into the silvery abyss, a clouded arm immerses me in a watery embrace. After we share solemn words of hope, the transparent limb reluctantly returns to its enchanting white sea of oneness, of being.

Brown cows with horns sharp and souls sweet join my third night. By my side they stay like fearless guardians; motherly and protective. Tonight I am not alone; maybe I never was! Lying next to the crackling fire, dinner cooks slowly to the soft smell of lavender crushed under their bellies. I watch stories born from fiery flames; messages flickering in endless shapes that seem to lick the galaxy above. Simplicity, spirals, and time unknown strain away my worries, as questions slip in and out of my mind. Does a star strive to shine or fade into dust? Why did humans evolve to worry so much about hours they never had and never lost? Cow bells ring around me like nursery rhymes playing in a black crystal room of endless twinkling lights. The wondrous Milky Way guides our dreams as Sirius watches over us; cows and person curled up like an ancient herd, each one protecting the other.

Morning air pulsates with hoverflies that wake me from my slumber. The call of another day scented by rising dew. With no desire for tomorrow or yesterday, I could live like this forever, a true human for once flowing with the tide of time; never beginning, never ending. Existence in harmonious swirling rhythm.

The Present

The experience of internal time is of one present moment after another—the present moment of *now* where all life unfolds. In the present moment a numinous state is available to us where we step out of illusion and touch the divine. Eckhart Tolle, in his highly acclaimed book *The Power of Now,* writes:

> The more you are focused on time—past and future—the more you miss the Now, the most precious thing there is. Why is it the most precious thing? Firstly, because it is the only thing. It's all there is. The eternal present is the space within which your whole life unfolds, the one factor that remains constant. Life is now. There was never a time when your life was *not* now, nor will there ever be. Secondly, the Now is the only point that can take you beyond the limited confines of the mind. It is your only point of access into the timeless and formless realm of Being.

In this Now moment there is no such thing as not enough time—there is all the time in the world. The Now knows no quantity; Now only knows quality. If you can engage with Nature in total presence for five minutes, this is far more valuable than an hour spent in Nature in a state of distraction.

Athletes speak of a particular state of focus commonly referred to as the zone, where their attention is devoted entirely to the present moment and the task at hand without distraction. When you can be in the zone while in Nature, you become single-mindedly involved in whatever you are exploring while enjoying yourself the whole time. Another term for this state is being in the flow. The flow state is now considered a branch of the field of positive psychology that focuses on understanding the qualities and behaviors that lead to a life of well-being and happiness.

Jeanne Nakamura and Mihaly Csíkszentmihályi, cofounders of positive psychology, identify the following six factors as encompassing an experience of flow. The following list is from the book *Handbook of Positive Psychology*.

- Intense and focused concentration on what one is doing in the present moment
- Merging of action and awareness
- Loss of reflective self-consciousness (i.e., loss of awareness of oneself as a social actor)
- A sense that one can control one's actions; that is, a sense that one can in principle deal with the situation because one knows how to respond to whatever happens next
- Distortion of temporal experience (typically, a sense that time has passed faster than normal)
- Experience of the activity as intrinsically rewarding, such that often the end goal is just an excuse for the process

In my own experience, especially when engaging with plants, I find that being present is the doorway to communion. In the Now moment, where being in the flow is possible, there is no distraction. I am interwoven in the vast web of All That Is where there is no separation. My student Heather shares an experience of being in the Now moment.

At my current place (fifty acres of bush in the country where the owners see themselves as stewards, do not cut anything, do not farm, only mow around the house, that's it, and the matron of the land talks to plant spirits), I love the quality of quiet here, no internet vibrations, no city noises with very few people around. I can get deep sleep here. I notice patterns and shapes of things when I walk around outside. I am more open to stopping and being in the moment and open to noticing messages, clues, just letting things grab my attention. I have less mental clutter so I'm more open. Every time I go outside I have a spiritual experience.

The stewards of this land have an old gravel pit, which is sort of grown over now, with big trees up on the ridge above it. One entrance they call the womb. There is a firepit with four directions marked where there used to be a sweat lodge and spiritual practice is still performed here. I enter and feel the vibration intensify in me—my breathing changes and becomes rapid. I feel like crying as I become aware of how much I have to let go of. Sumac, Cottonwood (Poplar), Birch, and Maple are here. Are they here to help me?

Stepping through the Portal

One of the profound aspects of being in alignment with internal time, where the Now moment is so readily available, is accessing multidimensional realities. The first three dimensions compose the realm of spatiality—height, length, and width—and make up our everyday waking life. Einstein, through his theory of general relativity, introduced time as the fourth dimension and combined the first four dimensions into the concept of the fabric of space-time, or the space-time continuum. This dimension of time includes the mind, body, and the external world as well as the inner realms, so there is the possibility to experience both external and internal time in the fourth dimension. There are other dimensions beyond these. String theory in quantum physics suggests there may be as many as ten dimensions. However, the one that holds the most potential for raising consciousness is the fifth dimension. GurujiMa, in his online article "Moving Toward Fifth-Dimensional Awareness," suggests that "the fifth dimension . . . is not spatial nor is it temporal. Rather, it is a dimension that brings space-time into relationship with the timeless and eternal. Fifth-dimensional 'space' and the awareness that accompanies it creates a movement of consciousness rather than a movement on the physical plane. This movement allows us to begin to perceive the unity of life and matter because we are moving within a higher plane."

When we are able to shift into the higher plane of the fifth dimension we have a greater view of the nature of reality and the veil of separateness that we hold to be true when we operate in the third dimension falls away. Medical intuitive Kimberley Meredith in her book *Awakening to the 5th Dimension* theorizes that "the 5th dimension is pure love, pure light, unconditional forgiveness, unconditional acceptance, instant manifestation, unlimited possibilities, beyond time and space, the God vibration, expansion, healing, ascension, oneness, and interconnectedness."

So how do you know if you are entering into the fifth dimension? Attorney Todd Skyler, in a Quora thread answering the question, "What are the signs that you are in the process of shifting from a third to a fifth dimensional state of consciousness?," suggested that "if you are beginning to see the world from an expanded state of awareness, you may very well be shifting from a third to a fifth dimensional state of consciousness." He further noted that "if you believe the following circumstances are arising in your life, you likely are in the midst of such a shift.

- You have reduced desire to fulfill egoic needs.
- You are far less judgmental of others as well as their life choices.
- You have an inherent trust that you are supported by God and your spiritual guides.
- You have a profound desire to limit your exposure to that which is toxic and counterproductive.
- You find yourself wanting to spend more time in nature."

This last item on his list is the key that unlocks the multidimensional portal. When you are immersed in Nature, totally present in the Now moment, you can step through a portal into the reality of the fifth dimension where you are able to exist outside the confines of external time and third-dimensional space. Within the fifth dimension, you are no longer separate from not only Nature, but also from the universe (or some would say the multiverse). In this reality you can experience the truth of your divinity within (immanence) and without (eminence) because *you are of the divine.* In nonspace and time you are engulfed in a tsunami of the miraculous where you ride the tidal wave of all possibilities.

The following story from Earth steward Mark describes his experience of stepping through the portal via the Now moment.

Reunion with the Divine

Today, I walked the edges and banks and through the shin- and knee-deep waters of a wild mountain stream that courses through the land I live upon. With a couple of basic tools in hand for the purpose of doing some spring clearing, I first trimmed back deadwood along the banks before immersing in these waters flowing down from the still icy crevasses of the nearby north-facing higher elevations. While freeing up flats of gravel in the stream by moving larger rocks this way or that, clearing pieces of dam-creating fallen tree limbs, and hoeing free the gravel trapped by the newly created channels of increased current, a shift of perspective soon swirled within and without, and I felt a sense of being in the presence of what can be described as the mind of the stream, and her song, to which I joined.

While repositioning select stones among the strong moving waters of yesteryear's winter snowmelt, whose forces are strong enough to carry small boulders, the stream's and the stones' voices grew increasingly present, while the liquid, jeweled current circled around the minute indentations and protrusions of her banks. Her seasonally wider edges joined with her torso's core, as water changed course and changed flow in response to my maneuverings. Yet now it became "me" doing the responding, for she had me within her now, and there was a change of awareness from, "I'm doing this with these rocks and stones and gravels and flotsam so the water will do that," to the stream demonstrating that, "She'll do whatever she darn well pleases," yet she'll also move as I move.

What had been for me simply a task transformed into a relationship, a baptism provided by her holiness.

Some days later, while in the good company of my dear old friend Mark (whose namesake I share), who had traveled out from the Midwest to share some precious time with me while his wife was in Ireland and mine in the Pacific Northwest (both locations holding places and people dear to our wives), he and I were blessed with the most glorious of mid-April weather as we were in each other's company again in the nature-rich environs of my Vermont home. While here, it was at a Cedar-, Oak-, and Pine-lined shoreline that, for a whole afternoon and into sunset, we sat perched on a rocky slope among the trunks and fragrant aromas of these deciduous trees, water, and soil in the renewing air of springtime, to converse with one another, to reconnect and carry on.

As the day evolved we became incrementally enveloped by forces larger than our shared human history, he and I in the presence of dancing, sparkling light straight ahead, straight out from our sitting spots, the sun's light and warmth magnificent in this April day above the water's rippling, semicalm surface that swayed gently and subtly from west to northwest, left to right, standing still, then changing and moving back again. Through our voices stories came alive of former times and places of canoeing and boating, of fishing and camping, the memories hanging like a necklace of beads across the heart and breast, each bead a lake, a portage trail, a fire, or a witnessing of northern lights, each revisited memory existing independently in time and place yet strung together, while we sat with tea on Glen Lake's shores.

This lake has a way about her, a way of gifting daydreams, for her waters are glacier born and clear and relatively deep. Mountain ridges, composed primarily of slate, rise away in all directions. She lives unmotored upon and quiet, without houses, and stewarded wisely—words spoken softly, gently kayaked, and, above all, held sacred. Beaver, loon, osprey, and gull are home here among many other grand and minute beings. To consciously experience a Nature-provided daydream, one that is too often imperceptible in the whooshing white noise of modernity's everyday concerns, is not something to achieve or attain but is simply a blessing. This lake is good medicine, inviting reunion with the divine. We are to approach respectfully, gratefully, with an offering, if nothing more than just paying attention.

With time—and we did talk about big Time and little time there that day—and in our minds, both in our own way and our own experience, we incrementally drifted out from the shores of our own thoughts into Time. Not looking "at" the world around me (trees, clouds, sky, and water, surface ripples and silhouetted minnows, pond skimmers and Jesus bugs doing their high step), in an encroaching dreamy, siesta-like voice I spoke of the phenomenon of the soft-seeing way of being, the quieting and mellowing, the resultant soft, gentle, temporary erasure of thought and memory, and the sensual gravitational pull back into place within the vast web of life, with creation and infinity and mortality, and the growing awareness of being in the presence of all the faces and forces of the beings shaping Earth and sky.

And then, in that soft-vision state that is like a cat who gently comes to curl its furry self across one's chest and upon one's lap, the bright light of the late afternoon sun—still too brilliantly powerful to permit even the slightest of a direct, stealing glance—laid itself upon the surface of the water. Just as the most graceful New York City waiter would loft the purest of cloths into the air and let it settle on a table, the white light followed and was guided down upon the table of the lake's mirrored surface. I was no longer preoccupied observing this place in this time, no. "I" didn't exist separately any longer but instead had melted away, and my soul witnessed the nonhuman way of the Tao, the way of Now.

My words to my friend became less customary, and in that place between the world out there and the world inside, there was only the moment, the breath, the ripple, and the cloud, and I gazed out to see showering crystals of vertically descending sunlight in the air above the water. A field of dancing, sparkling diamonds fanned out on the water and light reflecting and ascending up off the lake from below, undulations of reflected light rising up to reach the undersides of the splayed Cedar leaves and Pine needles at the shoreline and also barbershop-poling up along the trees' barked trunks, carrying and shaping the sun's sweat beads of creation, an infusion of all that is holy on this Earth, in this Earth; a presence of being, of light both delivered from the sky and from the waters of Glen Lake, streaming down, across, and up into our bodies and into our minds and hearts. There upon the eastern shoreline, on a late day in spring, when the world and we were young again and also gracefully older and not alone, just at home.

SEVEN

Communication—Communion—Common Union

Communication comes in many forms and can be as simple as an experience that opens a particular door of perception, thus inviting connection.

Pam Montgomery

I have written and spoken about communication extensively, but it has thus far mostly been in reference to plants and trees. Now I want to broaden the scope and speak about communicating with Nature, beginning with the question: What does Nature's speech sound and feel like? And since we are a part of Nature, how do we remember the language of Nature?

Of all the forms of communication, verbal conveyance is the least effective. In face-to-face conversations, we prioritize verbal communication; however, body language and facial expressions can have an incredible impact on how information is interpreted. Albert Mehrabian, who studied body language, was the first person to publish research breaking down the components of a face-to-face conversation. In his article "Decoding of Inconsistent Communication," Mehrabian, along with colleague Morton Weiner, determined that when one person is faced with an inconsistent communication from another (for example, the other claims they are feeling fine while looking anxious, avoiding eye contact, or speaking with a falling tone), the receiver will pay the most attention to facial expression (55 percent) and vocal tone (38 percent), while relegating the literal meaning of the words (e.g., I'm fine) to 7 percent of their attention.

The problem that arises with verbal communication is that the words we use may carry one meaning, but we may actually feel something else. In contrast, the communication that occurs through our facial expressions, gestures, proximity, touch, eye contact, and vibratory resonance conveys our emotions, intentions, and positive and negative vibrations. When we are communicating with Nature, it is different from being in conversation with another human; however, the same theory applies, which is that verbal transmission—hearing words in our head, for example—is not always accurate, but the vibratory resonance we pick up on is *always* truthful and authentic. Because we are taught that our primary form of communication is through words, we tend to rely on them; however, when we perceive that we are hearing words from Nature, what we are actually receiving is a *translation* of the "touches," or vibratory resonance, from Nature. When our minds start to question what we are receiving—"oh, you're just making that up" or "that's just your imagination" (here, again, the implication is that imagination is a bad thing)—usually it means that we are misinterpreting the translation. Our work is to learn how to accurately translate the revelations from Nature. We absolutely have the ability to do this because we are of Nature, too; we've just forgotten the language of Nature. So it's time to get back on the horse and exercise our atrophied muscles of communication, bringing them back into shape so that the common union that comes about from communion with Nature is once again part of our second (or first) nature.

Monica Gagliano speaks of a bygone language that we all have access to. She references plants specifically here, but this certainly can be extended to all of Nature.

> When we learn to listen to plants without the need to hear them speak, a language that we have forgotten emerges; it is a language beyond words, one that does not wander or pretend or mislead. It is a language that conveys its rich and meaningful expression by bypassing the household of our mind and directly connecting one spirit to another.

Remembering and reawakening to the language of Nature is essential to taking up our rightful place as a part of Nature. Once we can effectively communicate with Nature our separateness dissolves.

Amateur naturalist Jim Cranford described Nature's language in a conversation with one of my students:

> In the natural world, beings don't communicate in the same way that we do. Communication is going on all the time, with everything around us. It's all part of an intelligent system. Therefore there's much more to it than just getting one feeling. Over time, our connection and relationship develops. Even if you live in the city, there is life growing all around you that is aware of you. They are aware of your attitude, your intention. Even our house plants are important to our well-being.

He goes on to share his thoughts on human's relationship with Nature:

> We think that we domesticated plants and animals, yet that's not how it works at all. We're not a big deal in the scheme of things, and it's important for us to be humble. When we finally come to the understanding that Nature is intelligent, we recognize that she is the supreme intelligence. We have no secrets that can be kept from her;

> we are part of [Nature]. What we build can be seen as no different to an ants' nest or a bird's nest. We are children of Nature, and we are doing her work as best we can under the circumstances.

Nature is communicating with us all the time by sending what I call touches. Touches are the little nudge, impression, feeling sense, intuition, or telepathic knowingness that comes across your awareness. This is an aspect of Nature's language that Nature is constantly using, attempting to get our attention. In response, we need to pay attention and then learn to interpret these touches from Nature.

One of my students, Jackie, describes how she communicates with Oak.

> *I have begun developing a relationship with Oak, and this beautiful White Oak in our front yard really wanted to be our Prayer Tree. This tree is also a great "sit spot" for me. Oak seems to lift burdens, helping me feel lighter when my load feels heavy. So far, I communicate with Oak by feeling energy enter my hands when held close to the trunk. Sometimes this energy can be strong enough to flow through my whole body and cause it to sway. This involuntary movement frightened me a bit when it first happened as I thought I might fall, and I would have if my feet had not been firmly rooted to the ground, holding me safely in place. I felt this when I asked Oak to be our Prayer Tree. I had pretty much asked every other tree in the yard, especially the ones with lower branches for hanging ties. But I took this as a strong yes, and now I cannot imagine any other tree holding prayers for us. I sometimes check the messages I get from Oak with a pendulum. This seems to help me feel more confident in honing new methods of communication. I sat for an hour to make ties and pray. It felt wonderful to make the time for so many prayers. As it turned out, the bark of the tree is perfect for affixing ties with small string. Interesting, too, that Oak has an "eye" that faces our front door—always keeping watch over our home.*

When I teach classes on plant communication, each participant has one plant they work with for the entire weekend. To determine which plant or tree to work or play with, I suggest to them that they move away from the anthropocentric headspace of being in control and to let the plant or tree choose them instead of the other way around. Nature is *always* aware of us and is observing us. As biophysicist Fritz-Albert Popp has demonstrated, plants (which are a predominant aspect of Nature) have a greater ability to fine-tune their resonance to us than we do to them. This means that the minute my students arrive, the plants and trees are observing them and deciding who they want to work with on this particular occasion. They are magnetizing their vibratory resonance to a particular person in order to catalyze a mutually beneficial alignment. When I send my students out to "be chosen," I instruct them to pay attention to which plant or tree they are attracted to since this is the way the plant manifests the magnetism it has already put into play. Perhaps you are attracted to a beautiful flower or the smell or the plant is waving wildly at you when there is no breeze. Sometimes there is a large plant or tree that gets your attention, but once you are with the plant you realize it is actually the tiny little one hugging the ground that is calling you.

When you begin to interpret the touches you receive from Nature, you put to use all of your inherent gifts of perception. Your ability to receive information through the five senses of touch, taste, smell, hearing, and seeing comes online, as does your intuition that is open to primal knowing, and you pay attention to the felt sensation of the vibratory resonance coming from Nature. You ride the waves of light and sound while you open your heart, the primary organ of perception, to activate your three centers of intelligence—the heart, the gut, and the brain.

Vibratory Resonance

Everything in existence has a vibratory resonance. Though objects appear to be solid, the molecules that make them up are in a constant state of vibration and carry a vibratory resonance. Our thoughts and emotions also carry a vibration. Places—either indoors or outdoors, such as a landscape, a house, a farm, a retreat center, the ocean or a church—and, of course, humans, animals, and all living beings carry a vibratory resonance. The speed or rate at which something vibrates is called frequency. Frequencies are made up of waves of differing length and height. So all vibrations have a frequency, as does energy. However, vibrations cannot produce energy because energy cannot be created or destroyed; it can only be converted from one form to another. As Riley Schatzle points out in his article about Nikola Tesla: "Everything in the universe is made up of energy, which vibrates at different frequencies." Perhaps this simple statement can bring clarity to the correlation of these three phenomena that are not usually visible to the naked eye.

Felt Sensation

In his book called *Focusing*, Eugene Gendlin wrote about felt sense and provided a process to help one to get in touch with one's emotions. I found in working with Gendlin's method that the experience of truly focusing on something was profound. I adapted his technique to working with plants and included it in my book *Plant Spirit Healing.* It is such an effective way of communicating because it puts us right in the middle of vibratory resonance, which is at the crux of understanding the language of Nature. I decided to reproduce that passage from my previous book, which I have adapted and expanded to work with all of Nature, not just plants.

When working with this particular technique, it is important to remember that felt sensation is a way to train oneself to tune into "feeling" vibratory resonance. However, the felt sensation is not the message; it is the messenger. And the information one receives, while quite useful, is still a translation. This can serve as an exercise for you to familiarize yourself with felt sensation.

Each aspect of Nature carries a different frequency to which we can attune. This attunement is received through the heart, which sends it to the brain that then registers it as a felt sense somewhere in the body. This initial feeling sensation is very important because it is what our bodies will remember on a cellular level. The light in the DNA of our cells receives this frequency in a holographic way, meaning that the entire picture of light, sound, and feeling sensation is recorded as a whole entity so that when we recall the felt sense, we actually receive the entire holographic imprint. A felt sense can come in many different forms like tightness, heat or cold, softness, expansion, contraction, itchiness, bursting, bubbling, and so on.

When you receive a felt sense, spend time getting very clear about the sensation and exactly how it feels. Take a moment to step away from the sensation and then come back. Is it still there? If not, tune in again to this aspect of Nature and feel again. When the felt sense returns in the same way, this is a good indication that you are receiving the frequency of this aspect of Nature.

Now you want to give the felt sense a word or image that exactly describes how it feels. You want to "get a handle on it," so the word or image is called the handle. Perhaps it is soft as butter or maybe a more accurate handle would be soft as silk. Go back and forth between the felt sense and the handle until it is exactly right.

Once you clearly have a handle on the felt sense, you can call it up at any time. It's like dialing the phone to make a direct connection. You ask to understand the felt sense; what is the meaning inherent in this frequency that you are receiving holographically? You may begin to feel an emotion like joy or grief, or you may receive images on your inner field of vision, or a memory may suddenly come up. As these images and sensations come alive, you send them back to this aspect of Nature, who confirms them in waves of yes vibrations or doesn't, helping you to discern and hone in on the impressions that are most aligned with the felt sense. Communication flows back and forth between you and this aspect of Nature until you begin to ride the same wavelength, vibrate at the same frequency, and resonate in harmony. Now you are able to perceive through this aspect of Nature because, even though you are still two entities, your energy fields have merged to the point of no separation.

Your body, in its original state, is all knowing, so you begin the felt sensation process here. The vibratory resonance of the aspect of Nature you are communing with mingles with your vibratory resonance, and you feel it as a sensation (not an emotion) somewhere in your body. You might experience the resonance as a body posture or a smell or even a sound, but in whatever way it comes, you must put a handle on it so you can call it back up. Once you have identified the handle, then you can engage your three centers of intelligence—heart, gut, and brain. An emotion will then begin to bubble up, which is not to be judged, just observed. When you begin to experience feelings both physically *and* emotionally, this is when you begin to come into resonance. At the same time, you notice everything that is going on around you—the sounds, the smells, the light, a bird or an airplane flying overhead. I call this entering the daydream of the aspect of Nature you are working with, where you pay attention to the Now moment that you are inhabiting together, and you become tuned to the experience of your Nature ally. All that you observe in the daydream of your Nature ally is woven into the fabric of who your Nature ally is and how they would like to be in relationship with you. Then you take all the pieces of this puzzle—the felt sensations, the heart's perceptions, the daydream observations—and put them together so that a full picture of your Nature ally emerges, and you step into common union where there is no separation between you and your ally.

Light

There would be no life on this planet as we know it if it weren't for light. Light is embodied by the element of fire, and the supreme fire is the sun. The other three foundational elements—water, earth, and air—are also necessary for life on Earth, but all life begins with fire and the light and heat it provides. Light is such a constant element of our everyday lives that we take it for granted and don't consciously realize how light is constantly communicating with us.

Light directs the daily, monthly, and seasonal rhythms in our lives. For example, in the north country where I live, light communicates to us by shortening the amount of daylight as the season changes from summer to fall. When the part of Earth where I am tilts toward the sun, the light becomes more prolonged and brings more heat and the arrival of summer, communicating a shift in the seasonal cycle again. You may say that this is so obvious, but what if this feature of our common union—the shared awareness that shorter days mean winter is coming—was not present? How would we know that it was time to prepare for winter?

Another huge aspect of light that we are completely dependent upon is the process of photosynthesis. In chapter 2, I discussed our symbiotic relationship with plants on a physical, emotional, mental, and spiritual level, and I want to reiterate some key points again. Through photosynthesis, plants, trees, algae, and some bacteria capture sunlight and use this energy to extract carbon dioxide from the atmosphere and then combine it with water to form sugars that make leaves, stalks, roots, seeds, and flowers, which contain starch, fat, and protein. The by-product of this process is oxygen. Plants, trees, and sea vegetables are the *only* source of oxygen on the planet, and they are able to produce this because of light. Also, plants are where all of our food comes from (or from an animal that ate a plant), and all of our body's tissues come from plants because we don't make tissue from sunlight, plants do. Plants make up 80 percent of all living organisms on the planet. This predominant, Earth element aspect of Nature communes with us daily in the form of food, medicine, breath, companionship, and guidance.

Plants are not the only organisms integrating light at a cellular level. In chapter 2, I also discussed biophotons, the particles of light that exist at the nucleus of all living cells. German biophysicist Fritz-Albert Popp writes in his paper "About the Coherence of Biophotons" that "the weak photon current from biological systems, which—as we know nowadays covers the whole spectral range at least from UV to infrared and which we call 'biophotons'—may well suffice to take the role of regulating the whole biochemistry and biology of life." These biophotons exist in *all* biological life—humans, animals, plants, and trees—and can generate coherence, creating a beam of light with other biophotons from other living organisms. This beam of light that carries biointelligence creates the possibility for us to communicate with animals, plants, and trees, as Popp says, "at the speed of light." We have the ability because of this light in our cells to be in common union *immediately* with other biological organisms.

Light carries vibration, which means that biophotons also carry a particular vibratory resonance. Popp has shown that in healthy human cells, the light particles are oscillating in an optimal coherent pattern, but if a cell becomes diseased then the "light goes awry." In his article "About the Coherence of Biophotons," Popp writes, "While normal tissue follows this optimization principle, tumor tissue has lost this capacity by a critical loss of coherence." Through a phenomenon called "photon sucking," biophotons from the cells of plants or other living beings could be shared with unhealthy cells to repattern or "reinform" the light particles of another organism toward health. I would call this opportunity to shift the light at the nucleus of your cell with the light from plants leading-edge health care. So the next time you sit with a plant and share vibratory resonance, remember that you have the potential to heal yourself at a very deep level.

The following is my student Lauren's story of healing in this way.

I struggled with health issues as a child and teenager, and instinctively rejecting the allopathic drugs being forced upon me, I found my way to natural healing, first studying herbs and then becoming a clinical herbalist As I worked to heal my digestive and hormonal imbalances as well as years of Lyme disease and inflammation, I worshipped at the altar of my countertop full of tinctures, teas, supplements, and superfoods. In my suffering I believed that if I could just find the right combination of things to put into my body I could heal. I experienced constant anxiety about eating food and being in environments that I felt to be toxic.

As I have come into deeper relationship with the spirits of the plants and my nature allies (particularly Mother Ocean), so many of the physical ailments that I have lived with have resolved themselves. This is because my relationship with the living world has led me down a path of uprooting and releasing the behavioral patterns, negative relationships, ancestral baggage, childhood traumas, birth wounds, and patriarchal oppression that have been held in my energetic field and manifesting on a physical level.

My supplement cabinet is now a bit dusty and stays untouched for days at a time. I grow a garden full of beautiful food and healing herbs, and I prioritize eating a clean, local diet of whole food, but I also occasionally eat ice cream or frozen pizza or enjoy a beer, and my body responds completely differently. This is because my soul has now found a home in my body, my feet are planted on Earth, and I am vibrating to frequencies of light that I receive from Nature, from Earth herself, from my healed ancestors, from the elements, and from Source. I am being nourished and held in alignment by this light on such deep levels that the physical nourishment aspect that I used to be obsessed with no longer carries as much as weight.

I recently noticed a significant shift in my body after I began deepening my relationship with Hawthorn. My lingering physical imbalances have been coming into equilibrium as Hawthorn opens my heart, clears limiting beliefs that are holding me back, and introduces me to new frequencies, including a pearly, opalescent, soft white light that envelops me in a soothing, energizing, and connective cloak. As I expand my energy field to be in communication with the land I live on, the ocean, the sky, and the spirit of the plants, I feel stabilized and woven into the web of life that is holding me. I know that this holding extends to before my birth and after my death, so in many moments of my life, I can live in freedom and the ecstatic bliss of being alive.

The increasingly present traits of ease, health, and grace that I walk with in my physical body feel like a by-product of being in deep connection with all of life, rather than like unachievable goals that I was taught to strive for by a culture that clings to the belief that we should all have these traits, and yet perpetuates a reality where we can't because we are disconnected and broken off from the very Source—all of life—by which we can embody them. Some days (not every day—I am still human), I feel beautiful in my being beyond what I ever could have imagined was possible.

I am grateful for the support that working with the plants as herbal preparations offers, but for me the difference between taking a dropperful of Hawthorn tincture two times a day, hoping that Hawthorn's physical properties will relieve an aspect of my suffering the way taking an aspirin will reduce a fever, and encountering Hawthorn as a seemingly limitless healing being who wraps me in eternal love and has the skills to heal me on deep levels and weave me back in the web of life, is like standing under a 50 watt light bulb compared to standing under the sun.

The way in which we primarily experience light is through color. Colors are waveforms that are made up by wavelength and frequency. The following chart adapted from Britannica.com illustrates the visible spectrum of light and the wavelength, frequency, and energy emitted by each color.

Range of the Visible Spectrum

Color	Wavelength	Frequency	Energy
Red	700	4.29	1.77
Red	650	4.6	21.91
Orange	600	5.00	2.06
Yellow	580	5.16	2.14
Green	550	5.45	2.25
Cyan	500	5.99	2.48
Blue	450	6.66	2.75
Violet	400	7.50	3.10

The color red, which has the longest wavelength and lowest frequency, also has the least amount of energy. Conversely, violet at the opposite end of the spectrum has the shortest wavelength with the highest frequency and the most energy.

We can consider the wavelength, frequency, and energy level that a particular color emits as a form of communication with us. We also carry emotional associations with particular colors, both personally and culturally, that are also a form of communication. Yellow is associated with cheerful, bright, and sunny feelings. Green is the color of Nature—of forests, fields, and plants—so, for me, green is very nurturing. The chakras, the seven energy centers of our body originally identified in ancient India, also have a color associated with each one. The heart chakra is green, which corroborates the sense of love and nurturance I feel from green. Pause for a moment: What color associations do you have? As you begin to pay attention to how color is intertwined with your emotional states, your memories, and nearly every aspect of your reality, you will come to realize that light is communicating with you constantly through the specific waveforms of color.

Shadow Soleil, an Earth steward and co-creative partnership advocate, shares her experience with biophotons (light) and biophonons (sound).

> *You spoke about the biophotons and biophonons inherent in plant cells and about their generous capacity to reeducate and encourage the integrity of our own cellular biovibrancy. After hearing that talk, I vowed to sit in deep meditation—every day at 3 p.m., for thirty days—next to, under, and with the lovely guardian Japanese Maple on my back patio. In this way, She would come to see me as trustworthy, reverent, and reliable and as wanting to enter into sublime communion with Her. The focus of the meditations often revolved around allowing our biophotons and biophonons to communicate and co-create, and we would both be embraced within our auric and acoustic sphere of sacred reciprocity, of synergistic divinity, with our own illumination and orchestration of love.*

Sound

It has been shown that light and sound are the foundational modes of communication in the biological world. Sound, like light, is composed of waveforms that are constantly interacting with our own vibrational frequency. Throughout the ages humans have understood and communicated—through myth, art, storytelling, and, more recently, modern science—that the universe originated with light and sound. Yet there is a debate as to whether sound or light came first. William Van Zyl brings clarity to this discussion in his book *The First Light and Sound in Our Universe: Contrasting Evolution and Creation*. The

Christian biblical creation version goes like this: "In the beginning was the Word [sound], and the Word was with God, and the Word was God" (John 1:1, NKJ). And next came: "And God said, 'Let there be light,' and there was light" (Genesis 1:3, NKV). In the Christian creation view, sound, or the word, was first, and then God spoke to bring forth light. Scientists who study evolution suggest the opposite. Van Zyl writes:

> According to the Big Bang Theory there was only light. Sound followed the explosion, but much later. It is a stark contrast. The question is: When was the first sound recorded on earth as we know it? Analyzing the Big Bang: there must have been "*sound and light*"—the initial explosion. However, one could reason, if there were any hearing animal or person out there, they would have heard it. Unfortunately, there was nobody—nothing—at that point in time.

In the biblical creation story, physical matter is created in six successive days by the words of God. The evolution narrative suggests that during the Big Bang there was a spark and an explosion followed, but the sound of the explosion didn't come until much later because sound travels at an exponentially slower speed than light. You could liken it during a thunderstorm, when we see lightning flash and then we hear thunder some time afterward, depending on how far away we are from the location of the lightning.

We experience sound in a myriad of ways, and each sound leaves an impression on us. Sound is constantly communicating with us, and we derive meaning from these communications. In Vermont, when I hear Canada geese honking high in the sky in their V formation, I know that winter is on the way. When I wake in early morning, just as the first light is blessing the sky, and I hear a symphony of birds singing their hearts out to find a mate to ensure the continuance of their song, I know this is a harbinger of spring. When I sit by Sweetwater stream listening to the cascading flow, I'm reminded that the sound of running water urges me to look at what resistance I have in my life that is not in service to me, and I feel great relief when I listen to the teaching of water to let go of what I no longer need.

On a few rare occasions I have received the great blessing of hearing the spirit of the forest sing. I described one such time in chapter 3, but the occurrence is worth looking into more deeply in view of my discussion of sound. I attribute the opportunity to hear the forest's song to being in the Now moment when we have access to other dimensional realities. When the portal opens and we step through into the unified field, we have access to the miraculous. And when we align our frequency (vibration) with that of the miraculous in Nature, anything is possible, including hearing the song of the forest. Miracles are available to us at any given moment when we choose to be in the Now moment.

Sensory Awareness

In my book *Plant Spirit Healing* I wrote the following about sensory awareness.

> We have evolved as sensual beings; our senses are one of the main ways in which we perceive the world in all its many guises; unfortunately our senses have become dulled by modern day life and all that comes with it—walking on pavement, smelling exhaust fumes, hearing motors running, tasting dead food, and seeing skylines of buildings. This dulling of the senses is one of the things that has caused us to be less connected and forgetful of our essential nature; yet when we are in Nature and walk barefoot, smell the sweetness of a flower, here the sound of running water, taste the burst of flavor in a wild plant, and see the outline of mountains, we come back to our true selves. Our cells wake up, are filled with recognition and remember that we, too, are a part of Nature, capable of engaging with Earth and all her beings in a deep and sensual way.

Dr. Zach Bush shares: "If we are not sensing the world around us, we will be unable to awaken to the deeper feelings of what it means to be alive." This statement rings true for me when I step outside and begin to observe all that is around me. My senses blend together—a symphony of wind in the trees, babbles in the brook, chirps of birds; riots of color in lime green leaves, sunbursts of Dandelions, pastels of water-splashed stones; the aromas of fresh Pine, sweet Lilac, moist Earth. I taste a feast of slippery Violet leaves, bitter Dandelion leaves, nourishing Chickweed, while touching it all with my heart and hand, running my fingers through the bed of moss, letting gravity hold me close to Earth, walking barefoot and receiving nanonutrients through the bottoms of my feet. Here with my Nature kin, I join in common union with all of life, and I thrive in my true essential nature.

One of my students, Leah, tells her story of communion with Nature by letting her inherent gifts come alive.

My Nature Allies

The ancient wooded path, sharp with afternoon summer light, is arduous and rewarding to walk as always. Draped with edgy stone shadows, lined regally with grass as beige and crispy as baled hay, adorned with Red Clover, Daisies, and Dandelion. Tiny stones massage my soles as chalk-white butterflies flutter like confetti over a bride. The sheer ascent of the path feels softened by Nature's beauty, sharing her language of love and encouragement for my sticky climb. I articulate gratitude from the depths of my solar plexus for her exquisiteness. Silence fills the air, broken only by a cascade of dusty leaves and the pleasant drip, drip of water. The stream meanders serenely; its sun-dried liquid navigating satin corners. Soothing birdsong echoes through rusty air, as a warm breeze strokes my bronzing skin. Beckoning it sweetly, a hoverfly lands on the tip of my outstretched hand. Eye to eye, I say: hello little friend, thank you for your company.

Ethereal space I sense amid branches and above mountain peaks framing a sapphire skyscape. Valleys hold memories of ancient ancestors' feet wandering the liminal land. In a sphere of speculative wonder, I ponder a time when we were one with Nature, in an affinity with the moon and movement. In this moment I feel as though I'm walking through a story, whispered by trees, rooted deep into my heart. Could Oak and Hawthorn really have liked Wordsworth's "Tintern Abbey" that I shared so full of animated energy last week, or the seashell and apple I offered with a bow of appreciation? I contemplate. Does my expression remind them of the pleasant sound of Celtic people's stories, Neolithic songs, or sweet woody music playing from the past to the dancing patter of shoeless feet? Maybe now, in this moment, the trees are sharing their stories, their memories, so selflessly and gratefully, in return with me. A friendship sprouting like no other as I initiate into the autumn season of my age. Like Nature herself, I'm awaiting to shed old ways in this passing, summery phase of midlife. Akin to me, she awaits the arrival of colorful wonder and the fall of needless detail. Innocent and bare again for winter's diamond arrival, holding tight only to water, love, mineral, and wisdom, I sense we've connected, mirroring our natural cycles as one.

Walking through rustic woodlands, I touch the twisted Chestnut bark of a dear old face, familiar from my hundred passing hikes. A glorious olden presence, holding dateless memoirs within. A delicate energy trembles through my fingers. When I embrace the tree, who I've fondly named Dungarven, the tree holds me. Trust built through time. No longer a stranger's greeting but communion and care, like being held in the arms of a loving grandparent. I wonder, Dungarven, how do you sense me? As I open my eyes, the vibrancy of color, the frequency of energy in a sense of wonderment shifts around me. Land brighter, clearer, an otherworldly paradise, right here, manifesting like a picture I'm being painted into. My breath taken away in awe, absorbed by effervescent jade, as I gasp the gift of life from plants into my walnut-like lungs. I trust I've been invited to peek into the veil in-between.

The tender touch of connectedness and kinship feels like whispers of unconditional love emanating energetically from soil-dwelling mycelium. With childlike curiosity, I know I am seeking what seeks me. So I continue to walk, smiling at the spicy smell of forest fruits, scented like hot sugary pies cooking on pear trees. Taking paths, long forgotten, into the wild unknown, guided fruitfully by nature into looming lands of so-called beasts, my Nature allies. Trusting trees and sacred stone will keep me safe; the footprint of a bear only reminds me where I am. I'm where the

Nature spirits roam, I think out loud, as I tenderly stroke the velvet surface of a Stinging Nettle in my forward pace. Following, without doubt, a single silver leaf shaking like an animate arrow and a vine blazing unearthly gold, luring me toward the distance. Intense rays of light illuminating rocks that guide me to the next stage. I recognize these signs are leading to secrets, unveiled before my eyes. Jumping in and out of sensory liminal space: dark and light, water and ground, glade and forest. Passing through a natural wild rose archway. The more I play, the more I believe, the more is revealed. A world existent all along, but unseen, unaccepted. Giving my human mind, dominated by modern distrust and fear, the affirmation to see, to sense, to believe, to feel safe. Everything I loved as a child—faun and fairy folk, tales told, talking with trees—is becoming nonfictional, a friend, not as imaginary as I was led to believe! After all, what is imagination? Thank you for trusting me with your secrets and for keeping mine safe within the caverns and grottoes of your miracle. With gratitude, my dear friend, have a good night. I'll see you again soon.

The Head versus the Heart

One of the biggest obstacles to communicating with Nature is the notion that the brain rules decision-making, communication, and greater understanding. The truth is the heart carries these gifts. However, people perpetually try to "figure out" how to communicate with Nature via the mind. Unfortunately, this is not how communication takes place, and this approach only leads to frustration and a lack of confidence. Eckhart Tolle says this of the mind:

> The mind is a superb instrument if used rightly. Used wrongly, however, it becomes very destructive. To put it more accurately, it is not so much that you use your mind wrongly, you usually don't use it at all. It uses you. This is the disease. You believe that you are your mind. This is the delusion the instrument has taken you over. Then the mind is using you. You are unconsciously identified with it, so you don't even know that you are its slave. It's almost as if you

> were possessed without knowing it, and so you take the possessing entity to be yourself. The beginning of freedom is the realization that you are not the possessing entity—the thinker. Knowing this enables you to observe the entity. The moment you start watching the thinker, a higher level of consciousness becomes activated. You then begin to realize that there is a vast realm of intelligence beyond thought, that thought is only a tiny aspect of that intelligence. You also realize that all things that truly matter—beauty, love, creativity, joy, inner peace—arise from beyond the mind. You begin to awaken. Identification with your mind creates an opaque screen of concepts, labels, images, words, judgments, and definitions that blocks all true relationship. It comes between you and yourself, between you and your fellow man and woman, between you and nature, between you and God. It's this screen of thought that creates the illusion of separateness, the illusion that there is you and a totally separate "other."

The following saying, likely paraphrased from Bob Samples's *The Metaphoric Mind*, suggests that the mind, in and of itself, is not all bad: "the intuitive mind is a sacred gift and the rational mind is a faithful servant. We have created a society that honors the servant and has forgotten the gift." When my friend and colleague Asia Suler approaches Nature or a plant, she informs the intuitive mind that she is now engaging with a sentient being, which allows the mind to be more receptive. I find that informing the heart of this is just as effective (assuming the heart is in coherence with the brain) because the heart then sends that message to the brain.

When engaging your heart instead of your head, ask your heart for guidance and then listen with big ears, meaning all of yourself. Open your heart to receive from Nature, paying attention or attending to your Nature ally. Nature is always reading your vibration and responding to the most predominant energy that you are emitting, so if you present with a lack of confidence in your communication, then the message you receive back may be confusing or undecipherable. On the other hand, if your communication is infused with love and a knowingness in your heart, gut, and brain that Nature is sentient and has been informing humans for 2.5 million years, then you can enter into communion and receive full guidance, healing, and cooperation from your Nature ally.

Ah, spring, my favorite time of year. My senses awaken in this burgeoning season as the warm sun touches my waiting skin that soaks up the radiance of the sun like a thirsty robin who becomes hoarse from singing his heart out. I smell the rich moist Earth as she releases the icy grip of winter, and I hear the rush of water cascading down the mountain as snow melts in the upper reaches.

And then I see her, the first Trillium, in all her majesty and burgundy beauty. I fall to my knees as my heart cracks wide open, giving thanks for this opportunity to be in the presence of such a majestic being. "Hello, sweet lady of the forest, it is me, Shares the Flower Song [I give her my Nature name so she knows who I truly am]. I'm so grateful to be with you on this fine spring day. Thank you for this opportunity to sit within your daydream. I will sit quietly with you and receive the touches of Nature that surrounds us, and if you might have something to share with me, I would be oh, so grateful to receive with an open heart and mind whatever that may be. In return I have brought you one of my handmade beads in hopes that it may please you."

As I sit quietly with Trillium, merging with her vibratory resonance, I receive bits of wisdom from her like pearls that I string on my Trillium necklace. When my time with Trillium comes to an end, I tell her how deeply grateful I am for this encounter and her willingness to share with me. I assure her that I will continue to be a spokesperson for the green beings and all of Nature as I continue on my beauty-way path. As I reluctantly leave this spot in the forest, I'm certain I noticed a nod from Trillium come my way, and I blow her a kiss as I head down the trail.

EIGHT
Joyful Encounter

Joy does not simply happen to us. We have to choose joy and keep choosing it every day.

Henri J. M. Nouwen

Biophilia

The innate attraction that humans have to Nature can be described as biophilia. The word *biophilia* means "love or affection (philia) for life (bio)," including all life, which is predominantly Nature. The word was first used by German psychologist Erich Fromm to describe a psychological orientation of being attracted to all that is alive and vital. Independently of Fromm, Edward O. Wilson, an American etymologist, also coined the term, using it to describe "the connections that human beings subconsciously seek with the rest of life."

Later, Wilson joined with Stephen R. Kellert to publish a collection of essays called *The Biophilia Hypothesis.* Their hypothesis asserts that humans' dependence on Nature "extends far beyond the simple issues of material and physical sustenance to encompass as well the human craving for aesthetic, intellectual, cognitive, and even spiritual meaning and satisfaction." In the periodical *Frontiers in Psychology*, Guiseppe Barbiero and Rita Berto expand the definition of biophilia, describing it as an evolutionary phenomenon.

Biophilia runs so deep in humans there is seemingly a genetic component. Kellert and Wilson assert that over the course of evolution, biophilia has become part of the human genotype through a process of coevolution of culture and genes and that individuals who are capable of becoming emotionally affiliated with their environment are bestowed with an advantage in terms of real fitness. This oceanic sense of bonding triggers the genetic memory markers that become what is known as instinct. We instinctively know that Nature is our kin and that the health of the natural world is intrinsic to our own health.

Biophilia clearly plays a part in our attraction to and love for Nature and maybe even Nature's love for us. Perhaps this is why the current climate crisis unfolding on a global scale becomes personal. Much like when a family member is being attacked or harmed, our immediate reaction is to come to their aid. Carl von Essen in his book *The Hunter's Trance* describes the necessary foundation of our relationship with Nature: "It is not enough to have an intellectual understanding of our relationship with nature and the burgeoning crisis of our biosphere. It requires an emotional involvement by humanity, a true love of the living world, and that comes from the spiritual experience of being at one with nature and all beings."

Plants and trees are an aspect of Nature that offer a particularly easy doorway through which we can experience biophilia. This biophilic connection comes from both our similarity and symbiosis with plants themselves and our long and close association with Nature throughout the emergence of our human species. Through this alliance, a type of bond occurs that is necessary for a healthy gaiacology (relationship with a living Earth). This bonding results in an emotional tie that opens our hearts to the reality of the love that Nature and plants have for us and that we have for them.

Experiencing this bond is a kind of homecoming. Some say that all sickness is homesickness. When we come home to the plants, Earth, and

Nature, we come home to ourselves and we heal. Deep healing can take place by being with what you love, what nurtures you, and what is in alignment with your true essential nature. We are affected by our genetically inherited love for plants, and we are healed by their love for us.

Heart Coherence

The key to experiencing joyful encounters and biophilia is being in a state of heart coherence. Our hearts beat constantly throughout our entire lives and produce an electromagnetic field with each contraction that extends and can be measured several feet outside the body. An important indicator of the health of the heart is its variability—the small fluctuations in the rate of the heartbeat, particularly in response to the external environment. Heart coherence can occur when the heart is operating in this healthy, variable state, which triggers the brain to become entrained with, or sync up to, the heart and other electromagnetic fields, creating a two-way communication flow and syncopation.

There are four primary emotional states that can enhance coherence: gratitude and appreciation, forgiveness, innocent perception otherwise known as nonjudgment, and offering care. The following excerpt from my book *Plant Spirit Healing* gives a thorough explanation of the first three emotional states, which provide positive impulses to the heart and amplify coherence.

> Showing appreciation elicits an immediate response in your body that lessens the stress response, causes entrainment with the brain, and affects the electromagnetic field around you with ordered coherence. It is easy to come into coherence while in a state of gratitude, because your heart responds immediately to any appreciation you can elicit even if it is not about the situation at hand. This causes your nervous system to naturally come into balance, lessens the burden of stress, and frees up energy to be available for creative outlet. Gratitude is a highly magnetic state and when one is in the state of gratitude, it is returned to you easily. Many religions suggest that

prayers are to be ones of gratitude instead of asking for this or that, and that by merely being sincerely grateful, blessings will come to you tenfold. Each morning as I begin my day I step outside and give thanks to the sun for its warm breath, give thanks to the earth for its sustenance, give thanks to the trees for the oxygen they provide, and give thanks to the pure water at Heart Spring that is the lifeblood of this land and provides the moisture needed to maintain my body. In each of these I recognize the face of the Holy—the spirit that vitalizes and gives me my very life—and I am grateful. To start the day in such a fashion sets the tone and allows appreciation to open the doors of the heart, letting the abundance of gratitude fill the vessel within you, easing challenges and freeing your energy to co-creatively engage fully with life.

Another positive impulse for the heart is that of innocent perception. This is the state of nonjudgment in which your view is like looking through the eyes of a child. With this perspective you see the world anew with freshness and are able to be present. Judgment is a part of our makeup, because in our evolutionary development humans have needed to make quick judgments, for example, to know how fast and far we needed to run to avoid the saber-toothed tiger. Judgment is a part of our fight or flight mechanism and helps keep us safe. However, the dangers that we now face are different from those ten thousand years ago. Constant judgment that creates a stress response is more harmful than helpful. Now it is more appropriate to let our hearts be discerning, allowing ourselves to make decisions that are more holistic, with less personal attachment, and to embrace other's [sic] opinions. It is then that we experience gentleness with ourselves as well as others. When I look at Dandelion and say that I know all there is to know about it, I'm not viewing it with innocent perception. Instead I am closing the door to any other possible experience or understanding about its character or healing qualities. My relationship does not continue to grow and the subtle nuances of Dandelion's gifts pass me by. When we sit in judgment we limit our experiences of life as well as our choices.

> Forgiveness is a quality that is harder to achieve than the previous two impulses but one that when accomplished is most rewarding. The difficulty lies in our responses to greed, betrayal, loss, shame, and dishonesty—actions that have hurt us to the point of locking down our hearts so as to never be hurt again. This vise on our heart causes it to atrophy, harden, and ooze power as if from a festering wound. It is also in this arena that we come up against some of our hardest challenges, ones that hold us in bondage and suck our life force. Forgiveness must begin with oneself before we can move on to others. If we can't forgive ourselves, our parents, our spouses, our neighbors, and our government then it is unlikely we could ever forgive the desperate act of a terrorist. And yet, it is this very forgiving that puts healing into motion by removing the shackles from the heart, allowing it to invigorate with coherence leading to compassion. If you continue hard-heartedness you are hurt not just once but repeatedly by the energy you expend holding on to your grudge. But forgiving, as Howard Martin so eloquently states, "releases you from the punishment of a self-made prison in which you are both the inmate and the jailer."

The fourth positive impulse of the heart is that of offering care. The heart's impulse to care for another is regenerating and uplifting to us personally and builds connection with others and with Nature. We can care for our elderly neighbor, and we can also care for our garden or local river or planet. Offering care creates an opportunity for our hearts to express our inherent gift of being good stewards in service to sweet Lady Gaia, our home.

Being in a state of heart coherence generates a consciousness grounded in abundance (love). Abundance means that there is more than enough. A lack of coherence generates poverty consciousness (fear) because without the inherent connection to all life that the heart experiences and informs the brain of when they are entrained, the brain cannot access this picture of the whole. Our brains operate by creating separate, linear thoughts, one after another, that, without guidance from the heart, can easily spiral into abstractions and lead us down the path of fear and isolation. The brain creates a belief that there is not enough and that we are alone, whereas the heart holds a holographic knowing that we are embedded into the vast and generous web of life and can never be alone.

Without loving relations, both human and nonhuman, humans fail to flourish. An article by Bridget Coila in the *Bridges* newsletter titled "The Influence of Touch on Child Development" states:

> Touch is essential for human survival; babies who are deprived of touch can fail to thrive, lose weight and even die. Babies and young children who do not get touched also have lower levels of growth hormone, so lack of touch can actually stunt a child's growth. The immune systems of children who are deprived of touch may also be weaker than those who receive plenty of physical affection; plenty of touch earlier in life can lead to physiological changes that might protect against later disease, including cardiovascular disease.

In opposition to the four primary emotional states that promote heart coherence, there are three primary emotional states that promote a lack of coherence. They are sentimentality, which is longing for what was in the past, anxiety, which is unidentifiable tension, and dread, which is apprehension or foreboding.

When our hearts are in a state of healthy heart rate variability, we are able to upregulate the release of oxytocin, a hormone produced both in the hypothalamus and heart. Oxytocin is fondly referred to as the love, or bonding, hormone and is responsible for social bonding, bonding with Nature, group memories, psychological stability, relaxation, trust, enhanced learning, and the diminishment of the sensation of pain from injury while speeding up the healing process. Oxytocin also initiates the "restorative response," which brings our whole being back to homeostasis and well-being. Psychotherapist Dorothy Mandel, Ph.D., shared with me an unpublished predissertation paper, "Activating the Restorative Response of the Heart," in which she writes about the effects of activating the heart's response.

Activating the heart's response shifts inner perceptions and body chemistry from an inner sense of threat to a sense of peace and safety. This shift rebalances the autonomic nervous system, broadens perceptions, and is a form of emotional and physiological self-regulation that actually allows us to respond more efficiently and effectively to environmental challenges. Recent research carried out at the Institute for Noetic Sciences suggests that these balanced frequencies, when present in one person, can affect others even at a distance. Emotional self-regulation is a simple lifestyle change that can increase intellectual function, emotional stability, physical health, intuitive knowing, and interpersonal harmony.

Oxytocin is the hormone via which we are hardwired to make connections and bond with all of life. The release of oxytocin is initiated by feelings of love and safety. When we are experiencing biophilia, which gives rise to love and a sense of being safely sustained by Nature, oxytocin naturally flows through our system and bonding occurs. Another other way that oxytocin is released is during the *aha!* or WOW moments of life. As mentioned previously, I refer to WOW moments as Wonder Opens Wide. These are the times when our breath is taken away by a burning red sunset or the heady aroma of a freshly opened jasmine flower, or when a hummingbird hovers in front of our face while its wings beat fifty to eighty times per second. In these WOW moments, we are like children filled with the wonderment of Nature. Our hearts crack open, oxytocin releases, and bonding occurs, and we make a deep connection with the sunset, the jasmine flower, and the hummingbird.

We are on a journey to Laguna Ojo de Liebre (which translates to Hare Eye Lagoon and is also known as Scammon's Lagoon, Scammon being a famous whaler), near the town of Guerrero Negro in the Mulegé Municipality of the Mexican state of Baja California Sur. The Pacific gray whales have made their annual migration to birth their calves. What a treat to have the opportunity to be up close to these massive mammals who weigh approximately ninety thousand pounds and live to be seventy-plus years old!

The boat we are on, called a ponga, *eases its way through the calm waters of the lagoon in hopes of an encounter with the large gray whales and their recently calved babies. As I look out across the lagoon, I can see spouts going off—over there and there, and, oh!, another one. It hits me that I am in the presence of hundreds, if not thousands, of some of the largest mammals on Earth, and they are all mamas who nurse their babies.*

We motor out quite a way into the lagoon. Our driver cuts the engine, and we sit quietly waiting. We are letting the whales come to us, not the other way around, as this is their terrain and we must be sensitive to their needs. I feel myself dropping into the zone of connection, opening my heart and inviting the big gray in. Within a short time, a very large being approaches and comes alongside the ponga. This great gray whale is longer than the boat but comes so close I can touch her barnacled back, and I am in awe. To my utter amazement, she rolls on her side and looks directly at me with her saucer-sized eye. Like the Eye of Horus that brings wholing and healing, I find myself in the Now moment of one of the deepest connections with any being I have ever experienced where there is only the interbeing of wholeness that I am a part of, and in this moment I am healed. As I dive into the depths of the great gray mama's eye, I feel her say to me, "I see you," and I know she sees me, *all of me, in my wholeness. The tears are running down my cheeks as the bonding of this WOW moment courses through my veins.*

Then, if this pinnacle weren't enough, an even more miraculous event occurs. Mama Gray lets her baby slide out from under her and then dives under her baby to lift her up so I can see her. It was as if she trusted me enough with her newborn and she was so proud of her she wanted me to be witness to the greatness she had achieved by birthing this beautiful being. As her baby gray perched on her back, I wept with admiration, gratitude, and some of the most profound love I have ever experienced. This was truly one of those rare moments I have always known are possible where I entered the big heart, the Holy Heart, where we are all connected, where there is no separation, where magic exists, where all life can thrive, and where love abounds in this world.

Let Work Become Play

Experience (and research, too) shows that play is an essential aspect of life that needs to be incorporated on a regular basis. I would suggest that a family that plays together stays together. This, of course, is a take-off on the original version of this saying, which is "a family that prays together stays together," implying the power of prayer. Perhaps we could combine these two together so that a family that prays and plays together stays together.

Years ago, my husband, Mark, and I were part of a school called Bolad's Kitchen, which was run by Martín Prechtel. We learned and experienced so much during the four years we participated in this school, but one of the big things that has stuck with me through all of these years is learning how to truly pray. The kinds of prayers I learned to make from Martín were, at a heart and soul level, so profound that I truly felt that I was speaking directly with the gods and goddesses of creation. When Martín prays it is with every fiber of his being. He used to say to us that he couldn't teach us to pray, but he certainly served as a good example of what praying looked, sounded, and felt like. When we spoke our prayers out loud—which we always did, lest the Holy might not hear us—Martín invited us to speak as eloquently as possible, "like honey rolling off your tongue." The act of praying has to do with "feeding" the Holy, so your prayer needs to be as inviting and luscious as possible. Our prayers became an artform and were creative, expressive, heartfelt, and also fun—a joyful act.

Since those days with Martín at Bolad's Kitchen, Mark and I have been in many ceremonies with others, solitary times around the fire or intimate moments where we are in Nature and are moved to pray together, and we always luxuriate in the bliss of having the blessed opportunity to speak directly to the Holy with honey rolling off our tongues. Sometimes I feel almost giddy with the joy and playfulness that praying can evoke. It is as if I am not only feeding the Holy but I'm feeding myself, too. In this precious act of being in deep gratitude for the gift of life and for the ever-present spirit that flows through this life, I am completely connected, and when I start my day in this fashion, making prayers, co-creation abounds.

I relish these wise words, whose concept originally came from German philosopher and psychologist Karl Groos, that have been passed down through the decades: "We don't stop playing because we grow old; we grow old because we stop playing." These days, as I find myself well into my eighth decade, these words ring true. I seek pleasure, fun, laughter, and WOW moments because it is what makes me *feel* good. There are so many benefits to playing that I feel that as I age this is one of the best ways to maintain my health. The online article "The Benefits of Play for Adults" provides the following list of the benefits of play.

- **Relieve stress.** Play is fun and can trigger the release of endorphins, the body's natural feel-good chemicals. Endorphins promote an overall sense of well-being and can even temporarily relieve pain.
- **Improve brain function.** Playing chess, completing puzzles, or pursuing other fun activities that challenge the brain can help prevent memory problems and improve brain function. The social interaction of playing with family and friends can also help ward off stress and depression.
- **Stimulate the mind and boost creativity.** Young children often learn best when they are playing—a principle that applies to adults, as well. You'll learn a new task better when it's fun and you're in a relaxed and playful mood. Play can also stimulate your imagination, helping you adapt and solve problems.
- **Improve relationships and your connection to others.** Sharing laughter and fun can foster empathy, compassion, trust, and intimacy with others. Play doesn't have to include a specific activity; it can also be a state of mind. Developing a playful nature

can help you loosen up in stressful situations, break the ice with strangers, make new friends, and form new business relationships.

- **Keep you feeling young and energetic.** Play can boost your energy and vitality and even improve your resistance to disease, helping you function at your best.

When engaging in co-creative partnership with Nature, I find that if we enter with the attitude that it will be hard work, then the belief that we can have fun starts to dissipate. In this state we think we have to "figure it out" with our head, which usually means we are missing the mark because joyful encounters with Nature happen via our heart, not our head. The following is Martha's account of her joyful encounter with Nature, tapping into her heart and being in prayerful gratitude.

For a while now, I have had a regular date with Nature, usually weekly, and I go alone. I look forward to these times as eagerly as I would anticipate a meeting with a lover. Today, I went to McDowell Creek Falls, about an hour from where I live, a place of magic and beauty. After I ate lunch, I felt called to do some reciprocity with Nature. I hadn't prepared any gifts in advance, but usually there are some small things I can use for offerings in my day pack. So I made bundles filled with Lavender flowers from my garden, using, with permission, Salmonberry leaves and grass for wrappers. As my hands worked on the little bundles, my heart began to open.

I remembered an interaction from last night that I needed to apologize to my partner for, and I had a moment of feeling bad about myself. Then I remembered that forgiveness is always an option; it starts from the center or self and moves outward like a ripple. Right away, I felt the spirits of the trees all around me, leaning in to hold me tenderly and calling me to be kind and forgiving with myself. Stepping into this forgiveness, I felt my heart open fully and my spirits soar. I sat with these beautiful and generous tree beings and these good feelings for a bit and sensed the expansive nature of forgiveness. As the feeling grew, I began to feel forgiveness for others in my life, from the micropersonal to the macropolitical. I hadn't consciously intended to use forgiveness as a common heart opener, but it worked wonderfully.

Then I engaged in reciprocity with three large elder trees—Hemlock, Fir, and Alder—and with the rushing creek. My prayers went like this:

> Beautiful Grandmother Hemlock, thank you. Thanks for the breath you share, for the fragrant aerosols you dispense on the wind that upregulate my immune system and reduce my stress. Thank you for being here a good long time already, showing me how to grow old in wisdom, grace, and beauty. Thank you for the inspiration, the peace, and the healing I receive in your presence. Thank you for holding me up in my times of need. May you and your descendants grow and prosper here always. I love you.

I've often wondered why being a good worker is considered such a worthy trait in our culture. Finally, I realized it has to do with production: the harder you work the more you produce, which means the more there is to sell and the more the consumer culture of capitalism thrives. The roots of this hard-work syndrome we are afflicted with in the United States began when the Puritans (English Protestants) arrived on the shores of this country. The Puritans believed that working hard guaranteed one's salvation and that God would be pleased if you worked hard. The phrase the "Puritan (or Protestant) work ethic" was coined in 1905 by Max Weber in his book *The Protestant Ethic and the Spirit of Capitalism.* Weber was the first to join religion and capitalism through the lens of sociology. However, many would argue that capitalism actually had its roots in pre-Reformation Catholic communities and that it comes from the idea that one's vocation or work in the world comes from a calling from God.

The reason I bring all this up is that one of the predominant things I hear when my students are, for example, learning to communicate with plants, is that if they work hard at it, they will succeed, and working hard is a serious endeavor that requires rigor, discipline, and time. This "work hard syndrome" is so deeply ingrained in our culture that, of course, one would think this is the route to achieving a desired outcome. I am not suggesting that working hard is inherently bad. What I'm encouraging is that work doesn't have to be serious and that it actually can be infused with joy and playfulness. One of the biggest proponents of this idea is Dr. Patch Adams, who wrote the book *Gesundheit.* Many of you may be familiar with his work, which became a major motion picture starring Robin Williams. Along with his lifetime's work of bringing transformation to the medical system, Patch was a professional clown, and he infused all he did with joy, fun, laughter, and playfulness. His words, "The most radical act anyone can commit is to be happy," express the importance of infusing our lives with joyful encounters.

As you walk through life, let your work become your play and let the beauty of each moment fill you with the gift of being alive.

NINE

Reclaiming Our Birthright as a Part of Nature

Nature is not a place to visit. It is home.

GARY SNYDER

Right Relationship

The nature of Nature is of spirit, which is the vital principle held to give life, and this vital principle carries "right relationship" at its core. John Humbach of Pace University School of Law offers a definition of right relationship:

> [R]ight relationships are relations in which each (or all) seek, without abandoning themselves, to be attentive and responsive to the needs and emotions of one another...That is, a relationship is not "right" if participants seek to overbear in power (oppress), to overreach in resources (exploit), or to mislead for selfish advantage (manipulate).

In his definition he is referring to people, but these principles certainly can be applied to the natural world, as well. To be in right relationship with Nature, we must stop the oppression we perpetrate by dumping plastic in the ocean, clear-cutting forests, and killing soil. We must stop our exploitation of Nature through treating it only as a commodity and resource instead of a life-giving matrix filled with sentient beings. To be in right relationship with that which gives us our life, we must stop our manipulation of water through building dams and our manipulation of plants through agricultural monocropping and the use of toxic chemicals, which kills pollinators.

The Cultural and Religious Roots of Right Relationship

In many different cultural and religious traditions, past and present, right relationship is a key concept, and these traditions offer guidance or model how to be in right relationship with Nature. In the Christian religion, the Bible instructs: "But now ask the beasts to teach you, the birds of the air to tell you. Or speak to the earth to instruct you, and the fish of the sea to inform you. Which of all these does not know that the hand of God has done this?" (Job 10:8–9).

Within the Celtic cosmology, right relationship with Nature is deeply embedded because the natural world was held to be inhabited by spirit. In Celtic Europe, a culture that originated in the sixth century BCE and whose traditions and beliefs are still alive in parts of Europe today (including Ireland, Scotland, and Wales), right relationship was not a secular endeavor but was considered a sacred act. The mountains, groves, and springs (wells) were considered holy, so to harm Nature in any way was considered to be a matter of life or death. If you harm Nature, you harm yourself.

In Zen Buddhism, the foundational concept of interbeing embraces the interdependence, interconnectedness, and interpenetration of all beings. Interbeing not only encompasses the connections between all life, and the ensuing relational dependence brought about by these connections, but also includes the concept of penetration or "the action or process of making a way through or into something," indicating that all beings are not only connected to but actually merged with all other beings, human and nonhuman. Right relationship, then, is essential as a core value of interbeing.

The concept of right relationship is inherent within Native American culture to the point at which it is embedded in their languages. The Lakota phrase *Mitakuye Oyas'in* is commonly translated as "all my relations." However, Lakota poet Taté Walker suggests that an alternative, deeper translation of this word is "I am in relationship with all." This word conveys a particular view of interconnectedness of all life. It is important to note that this expression is primarily spoken in prayer and ceremony and that to "use" it otherwise is a violation of the sacredness the word carries.

The Rights of Nature

There is a movement afoot to give rights to all plants and animals, to physical features, such as rivers and mountains, and to ecosystems, such as the Amazon basin. Behind the movement is the premise that all life has the right to thrive and that loving relationships, both human and nonhuman, are essential to all beings' health, well-being, and ability to flourish.

At a 2010 conference held in Bolivia, where 140 countries were represented, the "Universal Declaration of Rights of Mother Earth" was adopted. The declaration states:

> Mother Earth and all beings of which she is composed have the following inherent rights: the right to life and to exist; the right to be respected; the right to regenerate its biocapacity and to continue its vital cycles and processes free from human disruptions; the right to maintain its identity and integrity as a distinct, self-regulating and interrelated being; the right to water as a source of life; the right to clean air; the right to integral health; the right to be free from

> contamination, pollution and toxic or radioactive waste; the right to not have its genetic structure modified or disrupted in a manner that threatens its integrity or vital and healthy functioning; the right to full and prompt restoration for violations of the rights recognized in this Declaration caused by human activities; each being has the right to a place and to play its role in Mother Earth for her harmonious functioning; every being has the right to wellbeing and to live free from torture or cruel treatment by human beings.

Cormac Cullinan, from Enviropaedia, an online environmental encyclopedia, points out that "the Declaration is also a reflection of the legal philosophy known as 'Earth jurisprudence' which advocates an eco-centric approach to law and governance in order to ensure that human governance systems are consistent with natural systems of order." There have been many situations where the rights of Nature have been written into constitutions, most notably in Bolivia and Ecuador. In other regions, specific beings, such as the Whanganui River in New Zealand, have been given the legal status of personhood.

Pablo Solón, in his article "The Rights of Mother Earth," points out what sets apart the Universal Declaration of Rights of Mother Earth from other legal efforts to protect Nature: "This approach to Mother Earth rights sees that humans and nature are part of the Earth community and therefore we must see these rights as the rights of the whole and all its beings and not only of the nonhuman (nature) part." This distinction is key to a deeper understanding that we as humans must embrace, which is that *we are not separate from Nature but a part of Nature.* When we truly embrace our birthright as a part of Nature, change and transformation will accelerate, and we can become an Earth democracy, in the true sense of the word, where *all beings* have a right to thrive.

Creative Synergy

Synergy is defined as an "interaction or cooperation giving rise to a whole that is greater than the simple sum of its parts." It literally means "working together." Being creative has to do with using one's imagination, being innovative, and moving toward original thinking. So creative synergy is when we cooperate together to bring about something new or never thought of before as a solution or new creation. Julie Morley in her book *Future Sacred* writes:

> Creative synergy reinforces our kinship with the greater planetary ecology, the Earth node of cosmic process, by tuning us into the many other frequencies on that dazzling planetary spectrum of creativity. Our ability to be other-directed as individuals and as a species directly relates to our survival. Relationship (kinship, symbiosis, and synergy) forms a critical and sacred aspect of being and becoming together.

When reclaiming our birthright as a part of Nature, it is imperative that we move away from competition and embrace cooperation. As I discussed in chapter 2, what is emerging is that cooperation drives evolution, as Douglas Rushkoff indicates in his article, "Evolution Made Us Cooperative, Not Competitive." He writes:

> We've been conditioned to believe in the myth that evolution is about competition: the survival of the fittest. In this view, each creature struggles against all the others for scarce resources. Only the strongest ones survive to pass on their superior genes, while the weak deserve to lose and die out. But evolution is every bit as much about cooperation as competition. Our very cells are the result of an alliance billions of years ago between mitochondria and their hosts. Individuals and species flourish by evolving ways of supporting mutual survival.

When we step into cooperation, taking actions such as granting Nature her own rights becomes a rational and necessary step because we want our partners, both human and nonhuman, to be empowered in the partnership so that we can successfully create together.

In an interview with Julie Morley, ecology and evolutionary biology professor Bekoff notes that "*Future Sacred* [Morley's book] really made me think deeply about the importance of truly viewing ourselves as part of nature rather than apart from nature, and acting in ways that stress that we're all interconnected—humans, nonhumans, landscapes, and other nature—in 'sacred symbiosis.'" In the interview Morley refers to "the toxic narrative of Social Darwinism," where the notion of survival of the fittest is proving to have dire consequences.

> We are, of course, living in one such scenario when our imbalanced relations with each other and other species have created a pandemic. We find ourselves in what some call the Anthropocene—not the age of triumphant humans, but an age of reckoning, facing our human failures—the "nothing is sacred" perspective—that has affected planetary systems irreversibly.

When we engage in cooperation instead of competition we move beyond mere survival and create the conditions for a thriving ecosystem, where each participant has the authority to fulfill their role and is both making a contribution and being supported by the system. One of the most amazing examples of cooperation within an ecosystem comes from the work of Suzanne Simard, who wrote *Finding the Mother Tree.* In this groundbreaking book, Simard describes her research in the forests of the Pacific Northwest of Canada. What she found was that trees share and exchange food via underground mycorrhizal networks and that this fungal and root network is how the trees communicate with each other, even among different species. The Mother Tree that she refers to in her book is like the grand dame of the forest who watches over her offspring and cares for them. The impact of Simard's research is revolutionizing the way in which the scientific community, and all people who love and work with plants, view trees and their own relations within a forest setting.

The Mother Tree Project (MTP) describes itself as a "groundbreaking research initiative investigating forest renewal practices that aim to safeguard biodiversity, carbon storage, and forest regeneration."

Plate 1. *Wings of Sword and Light*, photo by Erin Summers.

Plate 2 (below). *Honoring Seeds*, photo by Leslie Silver.

Plate 3. Mother Waters, photo by author.

Plate 4 (below). Early morning in my sanctuary garden, photo by author.

Plate 5. North altar in my sanctuary garden, photo by author.

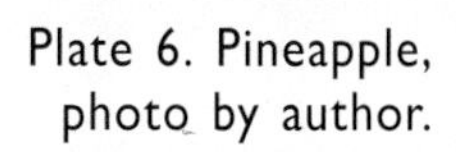

Plate 6. Pineapple, photo by author.

Plate 7. *Merging with Plants*, photo by Boris Austin.

Plate 8. Hawthorn mask, photo by author.

Plate 9. Labyrinth, photo by author.

Plate 10. Hawthorn gifts, photo by author.

Plate 11. *The Next Generation*,
photo by Shannon Sirjane.

Plate 12. *Dianne's Hawthorn Tree*, painting by Dianne Postnieks.

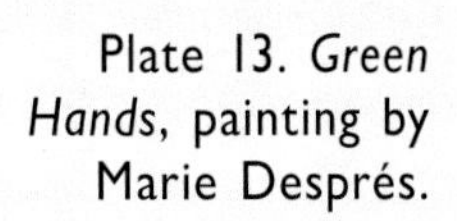

Plate 13. *Green Hands*, painting by Marie Després.

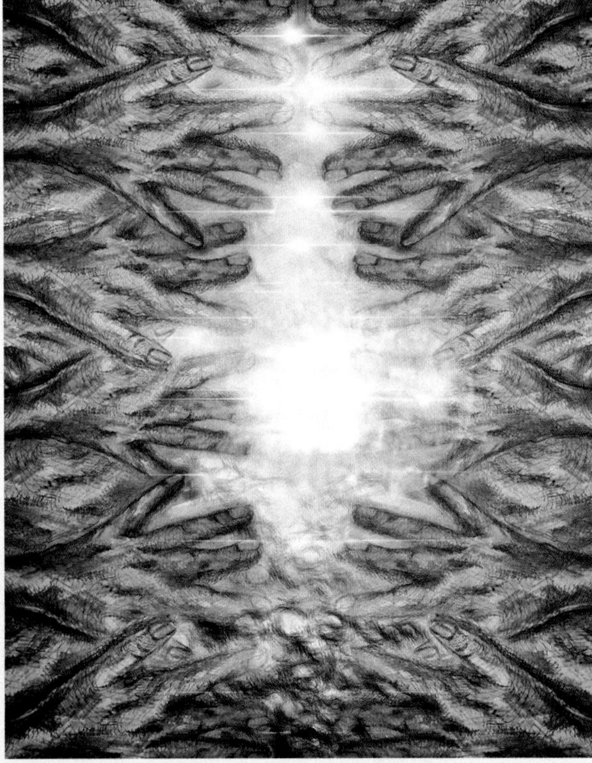

Plate 14. Full moon in Belize, photo by author.

Plate 15 (below). *Honoring White Pine*, painting by Lauren Valle.

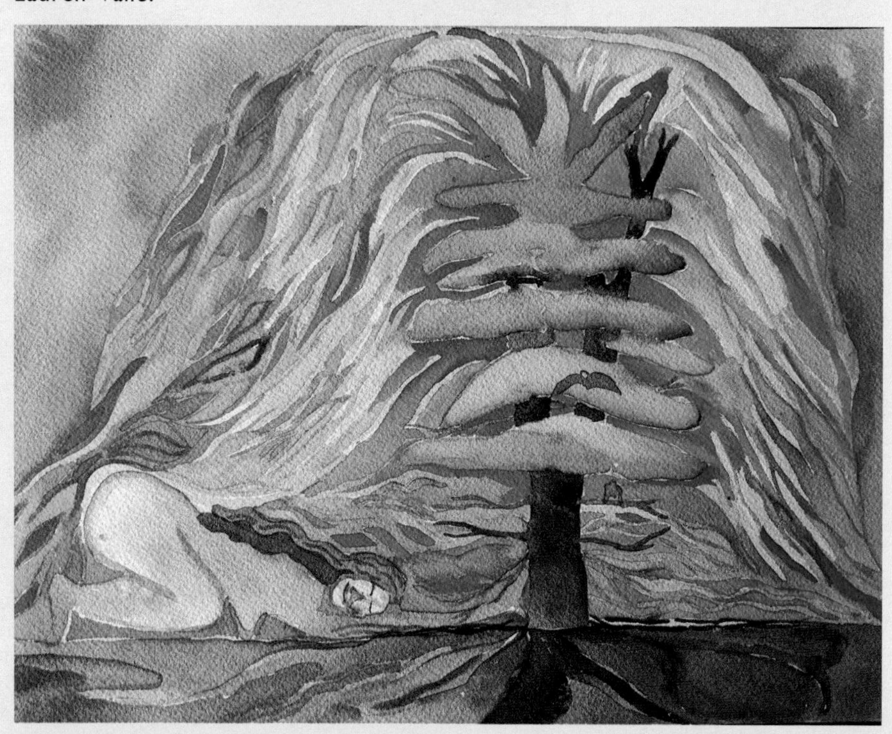

This project recognizes the importance of Mother Trees, and an article on their website describes how these Mother Trees function.

> Trees are part of a large, interconnected community interacting with their own and other species, including forming kin relationships with their genetic relatives. In mapping the fungal network, our research has shown that the biggest and oldest trees are the most connected nodes in the forest. These highly-connected hub trees, also known as Mother Trees, share their excess carbon and nitrogen through the mycorrhizal network with the understory seedlings, which can increase seedling survival. These Mother Trees in this way act as central hubs, communicating with the young seedlings around them. In a single forest, a Mother Tree can be connected to hundreds of other trees.

The work of the scientific community is making visible the previously invisible (to us) yet critical symbiotic relationships that are essential to the life of the forest, supporting what so many people have been sensing through the ages, despite our separation and amnesia—that all of life is connected. But what of the cooperation between humans and Nature? We are a part of Nature, so doesn't that mean that we, too, can cooperate with each other? Here is a story of Hawthorn's cooperation with me.

> *I sat near the Hawthorn tree during an initiation ceremony with Lady's Mantle. As I was contemplating what I had received from Lady's Mantle, my gaze wandered to Hawthorn, who had been planted at the edge of the labyrinth about twelve years prior when she was gifted to me by a group of my students. As I fondly admired how she had grown over the years, I commented that I had always wanted to facilitate an initiation with her as I had done in the UK several years ago, but I hadn't considered it because she had never flowered. Hawthorn immediately responded, "Well if you would do an initiation maybe I would flower." I was surprised to hear this, and I said, "So does that mean you want to be the initiator next year?" The response was a simple yes. The first Hawthorn initiation here at Sweetwater Sanctuary was about to take place in another week. With*

the emergence of spring and so much to do in the gardens and shifting into the new seasonal cycle, I hadn't visited Hawthorn until recently. I went to share with Hawthorn my ideas about the upcoming initiation and to hear what she had to say about how the initiation might proceed. As always, I approached Hawthorn with great reverence and gratitude for the gift of her beingness in my life. Then I saw them—little buds with white blossoms tightly folded, waiting to open when their time is right. I began to weep with the most magnificent joy. I said, "You are making flowers for the first time and are going to bloom!" She responded, "Well, you listened to me and here we are doing an initiation together."

This story, to me, is the ultimate example of co-creative partnership. I shared with Hawthorn what I needed, and she shared what she needed, and together we created the perfect manifestation. And this was just in preparation for the initiation. I could hardly wait to see what would happen during the initiation.

Being a Good Steward

One of the aspects of reclaiming our birthright as a part of Nature is to recognize our role of being a good steward who manages, protects, and serves with respect and trust and who honors the sacred responsibility of caring for Nature. As I discussed in chapter 1, being a good steward is the ultimate act in dynamic reciprocity. Nature gives us our very life so we respond with reciprocity—we give back by being a good steward.

Implied in being a good steward is taking action. This action can be making a daily prayer, engaging in a replanting project, attending a community meeting to speak about a project that is not in service to Nature or your community, or any type of action that has the well-being of Nature in mind. In taking this action it is important to allow yourself to be informed by Nature and not to assume you know what's best. Remember, this is a co-creative partnership where each partner has equal say in how to bring about a balanced manifestation where all life thrives.

One of the ways I have taken action to be a good steward is to create the Organization of Nature Evolutionaries, known as O.N.E. My original vision, over ten years ago, was to bring together a community of folks who not only cared about Earth and all her beings but who also recognized the sentience of Nature and that the vast intelligence inherent in Nature needed to have a voice at the table. As a Nature Evolutionary, I listened and gave voice to Nature. I was also observing that aspects of the environmental movement seemed lacking in a relationship with the sacred, and this was critical to bring in if we were to truly heal our separation from Nature. I became aware of the difference between the word *environment*, which means "that which surrounds," and *ecology*, which means "the study of home" and which is rooted in relationships. In an environmental movement, we are still separate from Nature, which is around us but is not integral, whereas an ecological movement is a coming home to ourselves and that which sustains us.

As a Nature Evolutionary I continually strive to shift the current paradigm to one that advocates for equal rights for Earth and all her beings. When we stand on equal ground, it is hard to forget the sacred that infuses Nature with the vital life-giving principle, the dynamic force that makes life live. In my work and play as a Nature Evolutionary, I strive to infuse the Green Renaissance with this very life-giving principle so that all the Earth work we do is done with the knowingness that the sacred is at the core.

Our vision at O.N.E. is to envision a future where people and Nature are co-creative partners and all life has the right to thrive. Our mission is to create educational opportunities in listening to and building relationships with the living Earth. Honoring our sacred connection with Nature is the foundation of our work.

Through this dedicated work with Nature, I have come to recognize the sacred in all life and am constantly reminded of the importance of honoring this sacredness through prayer, ceremony, and reciprocity. The Earth, the elements, and the plants give so much to me, and to be in right relationship, I must give back. So I give of myself in whatever way I can, but mostly I remember this sacred covenant humans have with

Gaia. When I remember the part of myself who never left the land—my indigenous self—then I step fully into my true essential self. As I embody my Nature self, I step into my primal knowingness that there is no separation between me and Earth, me and plants, me and water, me and air, me and the cosmos, me and all of life.

Ancestral Memory

All of our ancestors were people of the land at some point in time, even if it was thousands of years ago. We carry this ancestral memory in our blood and bones and at the nucleus of the cells within our very DNA. As we've touched on, this is possible through what is called epigenetics. The *Oxford English Dictionary* defines epigenetics as "the study of changes in organisms caused by modification of gene expression rather than alteration of the genetic code itself." The Greek prefix *epi-* in epigenetics means "upon," "on," "over," indicating genetic characteristics that are in addition to our traditional genes. An article in *Frontiers in Genetics* on epigenetic memory in mammals (by Migicovsky and Kovalchuk) described the molecular mechanism of epigenetic memory.

> Epigenetic information can be passed on from one generation to another via DNA methylation, histone modifications, and changes in small RNAs, a process called epigenetic memory. The ability of mammals to pass on epigenetic information to their progeny provides clear evidence that inheritance is not restricted to DNA sequence and epigenetics plays a key role in producing viable offspring.

In my own experience, I have found that when I visit the British Isles and Ireland, I have a distinct memory of walking the land. It's as if the memory is triggered by setting foot on the land, and as I walk, I have visions of a time past that marches across my radar screen. These visions are not part of my intellectual imaginings but emerge on a visceral level where my senses are engaged and my gut brain, or instinct, is fully alert. I recognize places, landscapes, and ceremonies and rituals whose memories are held in the land. I know this is the land of my ancestors, and their presence comes alive. These were people who lived close to Nature and were deeply connected to their landscape, where they honored the changing of the seasonal cycles, respected the animals they hunted for food, knew the wild plants as food and medicine, and lived with a knowingness that their very life depended upon the great Nature they lived within. When my ancestors immigrated to eastern Kentucky in the early 1800s, they brought their inherent kinship connection to Nature with them to this new land, which, to this day, lives on through me, as I embrace my birthright and thrive in my co-creative partnership with Nature.

Elyshia Gardner-Holliday, executive director of the Organization of Nature Evolutionaries, shares her story of reclaiming her ancestral memory:

Meet Me under the Yew

I can feel her bark through my shirt, imprinting her somehow familiar lines onto my skin. I am just breathing with her, taking in the scent and the medicine of this Elder and allowing my soft gaze to wander. To the left of me are the remnants of an ancient castle rising from the ravine floor where I sit. As my gaze arcs to the right, sunlight darts through the layers of the forest. Sometimes it reaches the leaf- and twig-laden floor, but often first reflects off the dense green life of the woodland. Bringing my gaze back to the center, to the small area that is brown, needle-covered soil, I feel another wave of intense gratitude and my chest expands and opens. The experience surges again through me. It etches itself into my mind, heart, and body, and I can hear my grandmothers whispering, "Now remember this . . ." I tilt my head to look up, admiring the branches simultaneously growing to Earth and sky. I feel at home, understood, and known.

It had only been a half hour that I'd been sitting at the base of this ancient Mother Yew Tree. When I first leaned against her, I had been plugged into another dimension. Over the two weeks I'd been here in Scotland, I'd had many experiences of being in other dimensions, but this experience with Mother Yew was different. There was no veil. It was intense and clear.

As soon as I had sat down with Mother Yew, a portal of black-blue light opened into the forest floor. At first, I rubbed my eyes in disbelief but the invitation was clear and insistent. I had come all the way from the United States to meet Yew. The spiraling light beckoned me.

Allowing my consciousness to move forward, I did not hesitate to enter the portal. I took note that the portal was going down into the Earth and the underworld. It is difficult to describe what happened on the other side of the door, but I can say that I found the voices of my ancestors, the voices that had called me here from thousands of miles away to be in the timeless space held by Yew. They knew me. Knew to call me. Knew that I would answer. But I knew nothing of them, and before this moment, I didn't even know I was missing them.

So, in this space held by Yew, my ancestors and I reencountered each other. I could touch them. The energetic strands of ancestral memory intertwined with the helical strands of my genetic code, reconnecting and healing the sacred line broken by generations of war, grief, violence, oppression, and colonization. For the first time, I experienced a felt understanding of lineage.

Finding myself in the underworld of the Yew was not a happenstance. Looking back, the journey had been weaving itself into being for more than two years. Having had a rich, Earth-based spiritual life for over two decades, in the past couple of years a deep grief had opened in me that felt like an unnavigable chasm. For decades I had been focusing on the lineage of my spirit and was completely oblivious to, and uninterested in, my human lineage. This path helped me really focus on spiritual development and worked well for me . . . until it didn't.

Beginning this period of grief, I found myself at a place that I think of as the point of convergence: this moment where the lineages of our spirit and our human ancestors need to be connected for our continued growth. When I arrived at this place of possible confluence, I didn't know my ancestors or really who they even were. For the first time, I could feel their absence and the hole it left in me. The intense loss I felt gave way to an ocean of tears, wild grief, and hopelessness.

The truth of my ancestral disconnection and the broken heartedness I felt created a fertile space. Even though the bulk of my ancestral memory may have been lost to time, distance, and cultural repression, the memories encoded into the sacred strands of DNA were still continually informing my body. I wanted to retrieve those memories. I needed to. We needed to. Within the European diaspora, our families have moved around Earth carrying a way of being that puts resources over relationships. This way of being has pushed people out of balance with Nature. To become good partners, relatives, and kin, we need to recover our own ancestral memories once again.

Months before this awakening, a friend had asked me if I would host a community discussion about decolonization. Through dreams, guidance, and remembering, the community discussion we had planned led to an ancestral gratitude ritual. Ultimately, this led to the creation of a circle we began facilitating that guided people through an eight-month arc of ancestral connection, story, and working with historical harm. Although my connection with my ancestors was strengthening, there was still this feeling of a barrier between us. The longing for the Earth wisdom of my ancestors intensified.

Soon after we had begun our second year of the ancestral memory circle, I heard a voice speaking to me from far away. It said, "Your memories are not lost. Your ancestors stored them for you." I couldn't believe my ears, but the voice repeated the message. Relief, joy, sunshine, and gratefulness washed over me. My heart, my blood, my spirit knew the truth of the simple words. Over the next year and a half, I continued to hear the message, "Your memories are not lost. Your ancestors stored them for you." Just hearing these two simple sentences was so strengthening, and each time the message came, a feeling or an image would also come. Eventually, I came to know that the being speaking to me was the ancient Yew, a tree elder I had never met and knew very little about. With some research, I discovered that Yew trees hail from a primordial time on Earth and that even today there are ancient Yews alive estimated to be thousands of years old. Many of these Yew elders were growing in Scotland, a place where I have many ancestors.

And so here I sat in Scotland, in the embrace of Mother Yew, the journey of the last couple of years culminating under her branches. I was immersed in the previously elusive confluence of my lineages. The chasm and disconnect I had experienced for years were now gone. There was a feeling of wholeness, and I could see my ancestors in a long line at my back. Very close to my right shoulder was a grandfather, and on my left, a grandmother. They were strong and sure, knew their place in the web of life, and they were with me. I thought, "Oh . . . this really is just how we are meant to be. This is part of our sacred human design." I had just been realigned to an important part of being an authentic, true human being.

And all of this was made possible by the partnership of Yew—first with my ancestors and now continued with me. Smelling the rich soil of the forest, seeing her long branches reaching from heaven to Earth, and aware of her bark pressing into my skin, I was overcome with immense gratitude. Honoring Yew's wisdom, long vision, and massive generosity, my heart overflowed with respect and love.

Through our ongoing relationship Yew continues to guide me in recovering ancestral memory, and I continue to be awed and inspired by this ancient Earth elder. I am grateful that our ancestors stored their memories for us, and grateful to Yew and other tree elders that have been the guardians of this precious wisdom through time.

PART THREE

When Plants Become Beloveds

Flowers enshrine my heart between their petals; that's why my heartbeats love them so much.

Munia Khan

TEN
Building a Co-Creative Partnership with Plants

The secrets are in the plants. To elicit them you have to love them enough.

George Washington Carver

In order for plants or trees to become your beloveds, you must build your relationship with them to the point at which you not only become partners but you fall deeply in love with each other. After falling in love, a union develops, where instead of being in love you *are love* with each other. To "be love" is a state of being that requires constant presence where, like a state of grace, you are influenced by the divine. The following are ways to build your relationship into a beloved co-creative partnership.

Breath

As I have emphasized throughout this book, we are incredibly symbiotically related to plants. One of the primary ways we are in relationship is via our exchange of breath. Plants, trees, and sea vegetables create all the oxygen on this planet, which is a by-product of photosynthesis. We breathe in oxygen and breathe out carbon dioxide, while plants breathe in carbon dioxide and breathe out oxygen. We literally exchange breath with the green beings in every moment of our day. Of course, because we breathe all the time, we take for granted the reality of this relationship with plants. Breathing is controlled by the autonomic nervous system (the aspect of the nervous system that regulates automatic functions, such as digestion, breathing, and heart rate), so we don't have to think every time we take a breath. However, when we do bring breathing, and where it originates from, into conscious awareness, we realize that we are deeply related to plants because our out-breath is the plants' in-breath and our in-breath is the plants' out-breath. When we are actually present with this exchange, the intimacy that is possible to share with a plant—through the generally mundane act of breathing—is profound. We begin building our co-creative partnership at this fundamental level.

Picture a favorite plant or tree. Breathe deeply and say to yourself, "I am the breath of another being." How do you feel? Does this awareness about the origins of your breath shift how you relate to plants?

Courtship through Reciprocity

When we court another, we are seeking their affection or we desire to attract them to us. The difference in working with plants is that they already have chosen you to be their ally. Being chosen by a plant means that you become attracted to a particular plant because it has already fine-tuned its resonance to you. As noted in chapters 2 and 7, this resonance matching was demonstrated by biophysicist Fritz-Albert Popp in his groundbreaking work with biophotons and plants. In essence, the plant begins the courtship by attracting you to it and drawing you in. Then, your part of the courtship begins when you enter into a state of gratitude and the desire to give back swells as your heart opens to this green being.

When we give back to Nature and, more specifically, to plants, this has been mostly referred to as "making an offering." However, my teacher Martín Prechtel suggests that we actually "feed" the spirit of the plants, and this is how we help to keep them alive. This level of reciprocity requires that we engage in depth perception to divine what would nourish our newly emerging ally and our budding liaison. So, for example, we would not likely give the gift of a bouquet of flowers to a plant since this is giving the plant what it already has. Ideally, we create beauty with our own hands as a way to give to the spirit of our new plant ally. Plant spirits do not have hands with opposable thumbs with which to create beauty, so this form of reciprocity is very desirable, and giving handmade gifts is a high level of courtship. We ponder what we can create that might feed the plant or tree—perhaps a beautiful painting or a poem or a sculpted piece of clay or delicious food or a woven piece of cloth or a strand of beads. When we present our gift to our newly emerging plant ally, with the burgeoning promise of partnership, we speak as eloquently as possible, so that our words are like a sweet embrace.

Physical Knowing

Of course, we want to know as much about this plant or tree as possible, so we engage our senses of seeing, smelling, touching, tasting, and hearing. Our senses are how we experience the world around us and, as such, can provide an enormous amount of details, impressions, and messages. As we explore our plant ally, little by little, like brushstrokes added to canvas, a portrait begins to emerge. (For an in-depth description of sensual awareness with plants please refer to my book *Plant Spirit Healing*, pages 84–95.)

We can also learn about our plant ally by ingesting the plant, which is often referred to as dieting the plant. Dieting is about ingesting plants as a way to get to know them, which is different from taking a plant preparation for medicinal purposes. The form of dieting that I share here is also not the same as ingesting an elixir during a plant initiation, which I will speak of later in this chapter. When we bring plants or tree preparations into our bodies, we are allowing our inner landscape to recognize and engage with the chemical and vibratory qualities of the plant. The inner landscape of our many body systems become acquainted with the constituents of the plant, and thus we begin to recognize each other's signatures or blueprints.

There are various forms of ingestion, primarily through combining the plant with the mediums of water and alcohol. A water-based infusion is when a dried part of the plant—the root, leaf, flower, berry, or seed—is covered with boiling water and steeped for anywhere from a half hour to eight hours or longer. A hydrosol is distilled water, which is either made as the by-product of distilling the essential oil of the plant or is a simple distillation of the plant in water. To make a tea, boiling water is poured over a plant and left to steep for ten to twenty minutes. Any of these forms of water preparation can be ingested for dieting purposes. A tincture is an alcohol-based preparation that can be brandy, 80 or 100 proof vodka, or grain alcohol, which is 190 proof, or 95 percent alcohol. The type of alcohol chosen to make a tincture is determined by the type of plant, its water content, and other factors. The fresh plant material is steeped in the alcohol anywhere from two to six weeks. Water and alcohol extract different plant constituents. Some plants, such as those that are highly resinous, contain constituents that are mostly extracted by alcohol, whereas plants with large amounts of mucilage—like marshmallow, for example—are more soluble in water. Then there are flower essences, which have no chemical constituents at all but carry the vibratory resonance of the plant. Flower essences are what you would choose to diet with if the plant is not considered an ingestible plant.

The length of time you diet a plant is really up to you, the plant, and the situation you find yourself in. It could be for three days up to six weeks. Usually, a diet is based around a natural cycle, like a moon cycle (twenty-eight days), or a blood cycle (how long it takes to replace your blood cells) of four to six weeks.

In her book *Journeys with Plant Spirits*, plant spirit healing colleague and author Emma Fitchett (Farrell) says this about plant diets:

During any plant diet, we view everything that happens in our life at that time through the lens of the plant we are in a container with. When viewed from this perspective, our eyes and minds are opened within the dream of the plant, and the magic of life and nature becomes more apparent. This is not some fanciful dreaming but a tuning in to the inner realms of nature and the level of consciousness that the plant spirits exist on. This dimension of understanding overlaps and is reflected in normal life; we can therefore observe the plant spirits working through the occurrences and situations that arise.

Heart Knowing

Our heart is the primary organ of perception, which means it is the organ that does all the perceiving, receives the touches from Nature, and, ideally, makes the decisions about what action to take or how to emotionally respond to any situation or interaction with a plant. When the heart entrains (or tunes to the same vibration) with a plant, then deep communication can take place.

When we open our heart to the plants and trees, it is easy to appreciate the gifts they carry, even if they are poisonous plants or cause welts or some type of dermal reaction. If we approach a certain plant as if they are the enemy, then our relationship immediately comes to a halt and does not progress. When we set aside our judgment and proceed with innocent perception, in gratitude with an open heart, then we can see the gifts of a plant.

A plant that could potentially be one that people avoid is Wild Parsnip. Some people call this plant Poison Parsnip, which, of course, it is not. However, because the juice of this plant can cause a burn on your skin, it has gotten a negative reputation. One year, the field adjacent to my garden became full of Wild Parsnip, and I was very curious about this and why so much of it showed up. It was a particularly challenging garden year, with long dry spells and lots of insects that wanted to eat everything in sight. During my investigation of Wild Parsnip, I approached this formidable plant with an open heart and with sincerity asked to understand its gifts. Because it is called Wild Parsnip, my mind automatically assumed it was the forerunner to cultivated Parsnip. So I dug the root of Wild Parsnip to see how it tasted. Because Wild Parsnip is a biennial (a plant that has a two-year life cycle), I found that the roots tasted best at the end of the first year and in the very early spring of the second year. When I researched Wild Parsnip, I found that it is a native of Eurasia and was brought to America by the English because it was highly revered as an easy-to-grow, nutrient-rich root crop, which becomes naturalized. I can imagine that it served as a survival food and sustained settlers through long winters. Wild Parsnip's main gift is that during hard times, or tough gardening years, it can be relied upon to provide adequate nutrition while being fairly tasty. So, of course, it showed up during a difficult gardening year to share with me its gift. Now, Wild Parsnips grow in the adjacent field every year, and even though some may call it "invasive," I'm happy to know it is here in case we need to eat it. Who knows what the future may bring, but reliable Parsnips are here and I am grateful.

Our heart is a very good listener, and when we limit distractions, our heart can listen deeply. When we listen with big ears (all of our being), we unlock the door to an ancient well of listening. As we plumb the depths, we find our intuition, which has unlimited access to All That Is. Here, our inherent knowing is merged with the unified field of plant intelligence.

Learning to Receive and Identify Vibratory Resonance

It has been demonstrated that the foundational form of communication in the biological world occurs through vibratory resonance, which is expressed via light and sound. Working with this universal language takes some getting used to, as it is a bit different from our everyday human conversations, but once you engage with it, you will find that it is quite effective and will never lie to you (as people's words can). You can refer back to chapter 7 for the full discussion of this form of communication.

What I have found over the years is that what keeps you from remembering and engaging in the language of the plants (and all of Nature) is second-guessing yourself or asking yourself if you are making it up. Once you let go of this limiting belief, the predominant language of vibratory resonance becomes first nature, and the omnipresent vibrations around you are revealed as a dynamic interweaving of sentience.

The easiest way to begin receiving and identifying vibratory resonance with a particular plant or tree is through felt sensation. Once I have identified the felt sensation that I am receiving from a plant to the point that I can give it a "handle," I can then easily retrieve the sensation, which is the signature of the vibratory resonance of the plant, at a later time. Because I have established a handle, in any given moment I am able to call on the plant for purposes of communing, healing, guidance, or joyful encounter. Once you master felt sensation you'll find it is an excellent tool for building your relationship with a plant into a co-creative partnership.

Sitting with Pineapple in Belize, it is the first time I have seen the flower of Pineapple. I feel the sensation as an explosion in my third eye—a geometric burst. It's like a mandala that fills my third eye area with a force. The design of the flower is very precise in its form, similar to the sacred geometry of the Merkabah symbol that has two overlapping tetrahedrons. This felt sensation is an expression of the sacred geometry of the flower. Wow! I've never had this happen before. The name I give to my handle to call up the felt sensation again is "Mandala Burst." Thank you, Pineapple, for your exquisite beauty and for bringing the gifts of the Merkabah forward through your conformation—perfect balance of the trinity (see color plate 6).

Dream Journey

When we are building our relationships with plants and trees into co-creative partnerships, we experience them in the dreamtime. The concept of the dreamtime, which differs from nighttime dreaming, was popularized by A. P. Elkin, an Australian anthropologist, who wrote about Aboriginal Australian spirituality, which included the dreamtime, in his book *Aboriginal Men of High Degree*. One definition of the dreamtime, also referred to as the dreaming or *jukurrpa*, is "the relationship between people, plants, animals and the physical features of the land; the knowledge of how these relationships came to be, what they mean and how they need to be maintained in daily life and in ceremony." In this explanation we see the relational qualities of the dreamtime space that we can experience, and also how we may be guided to live harmoniously with *all* our relations, both in mundane time and ceremonial, or sacred, time.

The nonaboriginal understanding of the dreamtime, or the dream journey, is an experience of engaging outside normal time and space in a multidimensional reality, which is accessed through various methods. Journeying through daydreams is available during our waking hours, when we allow ourselves to have a foot in each world—one in the physical world we reside in and the other in the deeper simultaneous reality that exists within the unified field. Nighttime dreaming is what we are most familiar with, which is when we have seemingly fantastical or otherworldly experiences or scenarios while we are sleeping.

The shamanic dream journey is mostly accessed via drumming, rattling, or vocalization. When I facilitate a shamanic dream journey, through either drumming or rattling, everyone in the room who is journeying either sits or lies down on the floor with their eyes closed. I suggest that my students visit their wise person within. I guide them to a particular spot where they meet their wise woman or wise man within. I take them to this particular being because I know that *everyone* has a wise person to guide them through the dream landscape. When the journey begins, you go to the below direction deep in the Earth (and inside oneself) where you meet your wise one who takes you to where the spirit of the plant or tree lives. You then have time to spend with your plant or tree spirit to learn, engage, and sometimes merge with this being. When you return from the dream journey, you have deepened your relationship with your plant ally, furthering the development of your co-creative partnership. (You can read a more in-depth explanation of dream journeying in my book *Plant Spirit Healing*, pages 74–81.)

My student Samantha shares with us her shamanic dream journey with Bluebell.

> *I was taken to a primordial forest, filled with old-growth trees with a beautiful stream with a small waterfall cascading, bordered by luscious ferns. There was a feeling of coming home to joy and oneness. I found myself standing in a natural clearing, a secluded place, untouched by humans. The clearing was surrounded by flowers, this place felt familiar and safe, like a soothing balm upon my heart, mind, and spirit. Moving closer to the flowers, my eyes took a moment to come into focus, as my heart filled with joy and astonishment, as I recognized Bluebell as the plant ally choosing to work with me. Totally unexpected and undeniably perfect. Soaking in the details, I found myself connecting deeper. I asked Bluebell why she was choosing to be my plant ally at this time, and she communicated telepathically: "To help you find joy and beauty in the simple details and to open your connection to the faery realm."*

Samantha recounts afterward, "I've never met a wild bluebell in person before, it doesn't grow here in Australia. I'm not sure if it has a scent, but I could smell the most heavenly fragrance wafting all around me. Intoxicating my senses. There is no doubt in my mind that I was receiving a true vision. As well as the fragrance that engulfed my senses, I was given the name Hyacinth which confused me. My logical mind tried to dismiss this because I thought it was an entirely different plant than Bluebell. It wasn't until later that I researched Bluebell and discovered that it is a variety of Hyacinth and it has a fragrance exactly as I experienced! I'm feeling very blessed and grateful for this experience and look forward to connecting deeper with this beautiful plant ally."

In another journey Samantha's relationship with Bluebell deepens.

> *Bluebell shows me that it is safe to be my whole self, safe to speak my truth, safe to dance and sing with wild abandon, safe to celebrate life,*

knowing that she is with me, guiding my path. I am never alone. There is a profound warmth and inner glow within my heart as I connect with her heart—like an inner, delicate fire burning within a lantern that illuminates the path forward, whilst reclaiming all of the missing pieces of myself. The weight of sadness that has burdened me, like a heavy mantle that I've been wearing, is lifted from me. Tingles and warmth radiate from my heart and the sudden remembering of who I truly am and how beautiful life can actually be. Bluebell reveals a vision of me dancing within a circle of tall standing stones. At the base of each stone, bluebells are growing in abundance. I dance up to each standing stone, placing my hands on each in turn. With all of the stones activated, a portal opens. I stand in the center of the circle and raise my arms to the sky. A beam of radiant light floods down from above and simultaneously arises from beneath my feet, bathing me in a healing column of light. I experience such exquisite heart expansion; joy, love, and appreciation fills every cell of my being. Tears of joy are streaming. I feel illuminated from within, lighter, brighter, and more expansive, connected and whole.

Greenbreath Journey

As I've discussed already, one of the most foundational ways in which we bond with Nature is through our breath. Breathing is so basic that we don't give it much thought unless it becomes impaired. What if we were to bring our breathing into conscious awareness, realizing that without it we would be dead within a few minutes? If this consciousness led us to the source of our breath, the plant kingdom, surely our relationship to the plants and trees, which are made up of all the elements, earth, air, water, and fire, would change. We would realize that this inherent bond is one that is vital and that without it there would be no human or animal life.

In many traditions breath is more than oxygen flowing in and carbon dioxide flowing out. Breath also carries what the Asian Indians would call prana, the Chinese would call chi, or what I would call spirit. Spirit comes from the Latin word *spiritus*, which means "breath."

This prana, chi, or spirit is considered to be life force energy, that which enlivens each of our cells. The respiration process is a rhythmic in-breath or inspiration and out-breath or expiration, which Swami Rama explains in his book *Science of Breath*.

> When a person dies, the energy leaves. The body is still there, but the prana departs. Here we return to the matter of breath because breath is the vehicle for prana. When someone dies and the vital energy departs, we say that person has "expired." On the other hand, when someone experiences increased mental energy and creativity, we say that person is "inspired." We indicate through our language an intuitive recognition of the relationship between the inspiration, expiration, and the vital energy necessary for life and creativity.

When we are inspired we experience mental clarity, an abundance of physical energy and emotional well-being. With spirit, the vital principle held to give life, coursing through us, we become plugged into life and all its diversity. The trick to being inspired is to be conscious of the life force that rides on our breath, as an eagle rides the currents of the wind, or foam eyebrows ride the crest of an ocean wave, or light rides the horizon at daybreak. This life-giving breath is crucial to our well-being, and yet we take it for granted, as if it will always be there for us even if we don't give it any recognition. According to *Science of Breath*, breathing is the only physiological process that can create change—physically, emotionally, mentally, or spiritually—depending on whether it is involuntary or voluntary. Consciously shifting our breathing patterns can change not only physical functions but also emotions, thoughts, and even our personality. When we awaken to our breath and its inspiring qualities, our creative power within becomes available. Many methods of working with the breath (such as transformational breathwork, holotropic breathwork, and rebirthing, to mention a few) have been developed over the years to serve as a tool for moving through stuck patterns that no longer serve us, helping us reach our full potential.

The beauty of breathing is that with every inhalation there is an exhalation so that we have an opportunity to release carbon dioxide, a by-product of cellular exchange of oxygen. This release of carbon dioxide helps set the rhythm of our breath as well as remove what we might see as the spent fuel from the energizing process. As we release we also let live, as our carbon dioxide contributes to the plants' and trees' in-breath. This reciprocal exchange of breath with the green world creates an interdependent relationship that is a primary bond between humans and Nature.

When we consciously breathe with the green beings, our bond with Nature deepens, and we move beyond our mere physiological needs being met to the nurture of Nature. In my work with plant intelligence, I continue to seek ways to expand my relationship with plants and trees, and focused breathing, which I call Greenbreath, is a profound avenue for bonding not only with plants but with all of Nature.

In the early 2000s, I had the good fortune to participate in a transformational breathwork session with a lively group of women. My dear friend and colleague ALisa Starkweather had come to Sweetwater to teach a class in women's empowerment and lead the breathwork session. When I heard how long we were going to breathe together—over an hour—I was astounded, and I realized this would be an excellent opportunity to go further in my exploration of relationship with the green world. Since plants and trees are the source of my breath, it made sense to me to focus with a particular green friend. I was working with White Pine at the time, so I chose to be guided by White Pine throughout, while also focusing my breath with my budding tree ally. The breathwork was quite intense, and many of my mother issues surfaced as I worked to understand and heal them. So much happened in this first Greenbreath session as I moved through limiting beliefs from childhood, my ancestral lineage, and beyond, and I realized this was the next level of intimacy with the green nations. I cried, screamed, laughed, feared, and somehow came out the other side breathing with White Pine, one of my most potent personal allies. As I breathed, my hands began to tingle, while my whole body received a warm flush of radiant energy. I began to wonder if this is what photosynthesis feels like as I merged with White Pine to the point at which I was not certain who was breathing—me or White Pine. I felt a tingling in my spine, which began to move upward. By the time it reached my upper back, I was in a massive kundalini experience, with waves of ecstatic energy moving in a snake-like fashion up my spine as I reached an ecstatic bonding with White Pine much like an orgasm. I had always thought one might be able to have sex with a plant or tree, but I wasn't quite sure how. Now I knew. This experience was like none other with a plant or tree and was formative for the creation of the Greenbreath process that I now share with my students.

After many Greenbreath sessions with myself and others, I have come to realize that bonding with Nature in this way is the best therapy one could hope for. It moves you through obstacles and limiting beliefs, physical compromises and emotional traumas, and removes intrusive energies, while filling you full of your true essential nature so you can walk the path that is yours. A Greenbreath session begins by sitting with a plant who has chosen you or tapping into an existing plant nonlocally (across time and space) that is already known to you. You set an intention with this plant or tree to help you move through that which keeps you from being all you can possibly be, while at the same time committing yourself to being a lifelong ally with this plant or tree. In this way, the relationship becomes reciprocal instead of a one-way encounter.

Greenbreath is a form of journeying into the holographic world of spirit where, through one individual plant spirit, the whole of spirit can be experienced. Since our breath is of spirit, when we focus our breath in this way we have access to enormous amounts of Source energy, which carries a high wave vibration or frequency. We breathe in a wave-like pattern of equal inhalation and exhalation, with no pause between the two. This creates an activation that puts us in a positive or yes flow pattern of energy movement. The high frequency that this activation creates can dislodge static or stuck energy and long-held negative emotions. The ego breaks down so that the constant monologue of the "nag" no longer interferes with our healing process. As the ego dissolves, past traumas that we have clung to begin to move in a flowing pattern, and just as waves break upon the shore, old hurts and traumas crash upon the shore of our psyche, breaking into millions of fine particles only to be washed away by our focused exhalation. Spontaneous healing—physically, emotionally, mentally, and spiritually—can occur. New neural pathways can be laid in the brain so that exploration of unknown territory is possible. Our Greenbreath opens doorways that we cannot even imagine are possible. Light rides upon Greenbreath as coherence occurs between biophotons at the nucleus of our cells and the cells of our plant ally. This radiant beam of light reinforms the light patterns in our DNA toward balance and well-being. Our entire energy field becomes ensparkelated with crystal clarity as a portal opens and the vision of our life is revealed. You remember who you are, where you came from, and why you are here. Through the focus of Greenbreath not only does your relationship with your plant or tree ally deepen, but also your relationship to yourself, to Nature, and to all of life expands. On the surface these statements may seem grandiose, but as you listen to the stories of those who have experienced Greenbreath, you get a glimpse into the evolution of consciousness possible through this foundational way of bonding with Nature and plants.

Greenbreath almost always takes place indoors in a space where people feel safe. Participants arrange themselves, lying down comfortably on yoga mats or cushions on the floor, with enough space between them so that the facilitators can move around the room. The entire Greenbreath journey is accompanied by evocative music that sets a tone for each phase of breathing. We begin with nesting music that gives us time to settle into our spot and get used to the open-mouthed circular breathing of equal inhalation and exhalation with no pause. As you breathe deeply into your lower belly, a wave-like motion is created between the in-breath and out-breath as a rhythmic pattern emerges. Sometimes there is difficulty, initially, with open-mouthed breathing, and yet an open mouth allows for you to take in more oxygen, which, just as adding wood to a fire fuels the flame, helps to create high frequency activation. There is no right or wrong way to breathe, but there are ways to breathe that shatter limiting beliefs more quickly.

Once the rhythmic pattern of breath is established, the tempo of the music picks up and becomes dynamic and active, with deeper bass notes, which moves us into a space with the potential to experience all the resistance that comes forward, such as not being good enough, smart enough, or pretty enough, being bad, different, or fearful, or not trusting one's own experience as authentic. I ask the questions out loud: What is keeping you from your true essential nature? What is keeping you from a path of healing for yourself or others? This phase of confronting resistance can be very challenging, and yet it is crucial not to skip over, as this is exactly where we get stopped in our tracks.

Janet says, "What emerged during the release section was not wanting to feel stupid. I guess I come from a fairly high-pressured intellectual family where it's important to be smart. And as the youngest, I felt stupider than the rest. The grief that brought up touched a deep loneliness—essential loneliness. I wailed." Another student reports, "[It] made me think about being born into this world, and I felt like a spirit coming into this world—but against my will. I distinctly remember thinking, 'I don't want to be born into this world, with this painfully lonely life.' I began to weep, all the while sustaining the breathing." This sense of loneliness that both Janet and my other student express is the disconnection from Nature and the source of our sustenance, which has occurred for most humans. Even when we have a seemingly close and conscious relationship with Nature, there is a deep well of grief filled with all we have forgotten about our bond with Nature.

The next phase of music takes us into our ancestral lineage. Many of the burdens that we carry we inherit through our bloodlines. They are the things that our ancestors left unhealed: trauma, being persecuted or being persecutors, anger, grief, loneliness, and longings. In order to be all we possibly can be we must lay down our ancestral burdens and let those ancestors, who are able to, help us instead of hinder us. During this phase, we ask what these burdens are that have been creating such a weight—the ones that don't seem like ours but have been causing such pain. We are encouraged to lay them down now.

Jill says, "I started talking out loud to the ancestors amidst screaming, crying, spitting, and blowing—a lot of letting go. There was weird vibrating on the left side of my body. It felt like stuff coming out of my second and third chakras. I was seeing an area of light from me going out the window and stuff was going—ancestral junk was leaving and they were being freed." In one of my own Greenbreath experiences I deeply connected with my grandmother and all my grandmothers before her. They said, "We remember you," and at the same time, the plants were saying, "We remember you," so it was as if the plants were my ancestors too, and they all were here to guide and help me. Tim shares his experience of ancestors: "The music changed abruptly to a Scottish-sounding song, and all at once I was aware of being hollow again, like a conduit deep into Earth, through which all my ancestors passed as they came into this world. And I felt connected to my distant, ancient ancestors in a way I have never felt before." One of my teachers once said, "It's hard to know where you're going if you don't know where you came from." The memory of our bonding with Nature partly comes through our ancestors because at some point in time, all of us had Indigenous ancestors that were deeply connected to earth, air, fire, and water. Once we lay down their burdens we can access this memory.

Greenbreath music gives us one more opportunity to let go of what no longer serves us on our soul's path: we look our obstacles straight in the eye. We see them, know them, and let them go with every exhalation of Greenbreath, giving over all of our barriers to our plant ally to neutralize. We no longer need to live with fear, not enough, doubt, shame, lack of trust, or guilt. With the help of our plant or tree ally, we breathe through to the other side. Marjorie describes the last vestiges of obstacles leaving her: "The weight and burden of guilt I had heaped upon myself begins to lift as I identify how much of my thinking divides me—very counterproductive. I see a bloody stake in my heart. I am lying down and removing it with my right hand. As I hold the stake up, blood drips off and the hole seals up. I plant the stake in the ground, and a tree grows with deep long roots with a full green leafy top. The green extends to my heart as it soothes and gently strokes my heart."

Now that we have released resistance, burdens, and obstacles, the restorative process can begin as the Greenbreath music shifts away from a dissonant, driving beat to a lighter, more ethereal realm. Our plant and tree allies have been with us throughout the difficult phase of release, giving us their life-saving breath, and now they come to us, committed in their presence, as we begin the journey to balance, well-being, and restoration to our true essential natures. On this journey home to oneself and to Nature, the origin of our sustenance—breath—becomes a bridge to Source energy via the green beings. The light of Source energy rides on this Greenbreath and makes a rainbow bridge of light, as breath travels up our bodies, moving through each chakra—the color red seated in the root of self, orange engaging the hara as we encounter "other," yellow fueling the fire of will in the solar plexus, green surrounding the heart with love, blue speaking truth in the throat, indigo perceiving with inner and outer vision in the third eye, and purple crowning our connection to the divine.

Marjorie describes the interplay of her chakras: "Chakras are in motion with colors so clear, as a triangle is made with blue in my throat as I find my voice, yellow in my belly as I know my identity, and green in my heart as I embrace myself in love, starting at any one point on the triangle each positively affecting the others." As boundaries melt away, the expansiveness of Source is a bath of rainbow light and you realize this Source energy is brought to you by the plants and trees, especially your ally that you are exchanging your Greenbreath with. Tim says, "I began to feel my legs and arms tingling. And then I realized that my legs were *roots*—and the tingling was the feeling of my roots extending far beyond my legs and feet. The tingling in my arms were the branches and leaves extending outward into the sky above me. Words have no power to convey the wonder and awe of this experience. I was in absolute *rapture*." Barbara remembers Yucca who had attempted to contact her fifteen years prior: "As we traveled along through the music, I was enjoying myself immensely, and Yucca finally appeared. 'Remember me?' he said. I flashed back fifteen years and thought, 'Yes I do!' I cried out 'I'm sorry, I didn't know, I didn't understand back then.' The flowers of the Yucca were glowing so brightly, the filaments that pull from the sides of the leaves were curled into spirals and explosions of fireworks. My body rose from where I was lying and brought me to the beautiful white light of Source, of Creator, of God. There was no fear, only anticipation and joy. I was given Yucca's healing gifts with the familiar feeling of 'lightning through my hands.' And I cried in gratitude."

As Greenbreath continues, you realize that your plant ally is also breathing. You breathe in your plant, but it also breathes in you. You are the breath of another being—you are the breath of your plant. As you share breath in this way, through the exchange of oxygen and carbon dioxide, an undeniable merge between you and your plant or tree takes place. Tim shares, "Chicory and I were dancing a vow to each other—or I should say that we were getting married. There were no spoken words or promises. It was more of a clairsentience, or knowing that I was being ritually married to Chicory. I use the term *married* because it is the closest term I can think of, but it was not in a carnal or romantic sense. What I experienced was more like an initiation—a spiritual merging took place between us." As you merge with your plant through Greenbreath it moves through your entire body in wave-like motions bringing life-giving breath to each cell. You feel *alive*, and spontaneous healing takes place. Jennifer says, "I received a healing from Redwood. I birthed myself. My spine aligned differently, and I stood in a new way—in an old way—in my full power. Earth peace is right in front of me." When healing occurs, you can move fully into the dance of life that your plant ally offers. Catharine shares her experience: "It was a celebration, a dance to life and living, both beautiful and carefree. It was pure joy! Light! Spontaneous!" And Sara says, "Then I had to *move*. Inviting Goldenrod in made me want to dance. My middle wanted to move in fluid waves. My arms wanted to writhe like snakes. I wanted to invite golden light into my core . . . to strengthen, empower, and energize. Goldenrod inside me was wonderful. I loved the bonding" (see color plate 7).

With Greenbreath coursing through your veins, you begin to remember your birthright of being a part of Nature instead of separate from it. You remember that you *do* understand the language of plants and trees and that you are connected to all of life. You remember yourself, your own true nature, and why you are here. Janet says, "I remember sitting around a fire with a small band of people dressed in furs. We were singing in tonal variations as a group, and each of our tones created a vibrational resonance that wove into an overtone that was constant. It was like our tonal variations were the weave, and the consistent tone was the weft, and an entire fabric of resonance was being created. Inside this sound we could call upon animals, plants, stones, water, fire, air, anything in Nature. Our resonance entwined with different aspects of Nature so there was no separation. We *were* the animals, the plants, the stones, but what was so incredible about this was that *everyone* was capable of this merging. It seemed this was preshamanic, meaning before there was the need for shamans to intercede with or interpret the spirits of Nature or the unseen realms. This was before the people had forgotten."

Bringing Greenbreath into your heart, you feel how full it is with the love from your plant ally. Your plant ally is giving you its breath, its Greenbreath, to help you live. Your plant or tree ally got your attention. It chose you; it wanted *you* to be its ally. Feel the unconditional love of this as you move deeply into your heart. Your heart and plant ally know what is needed for your heart to heal. Let the healing begin.

Rhonda felt enormous pain, guilt, and suffering from the divorce with her husband. Even though she felt she had worked hard to overcome the struggle, he kept appearing in her personal healing work. She describes the healing of her heart through Greenbreath.

> *Rose, Tulsi, and Mugwort are in me, I can feel them as I dive down into the ocean as if I'm a sea animal. My husband's presence comes and we're dancing. Rose is in my heart, in his heart and is up in heaven holding us in a triangle. Tulsi comes in her deeply rich purple gown and is dancing and swirling with him. Then he and I are dancing on a dock beside the ocean, and rose petals are falling from the sky all around us. Then we are standing outside of his business, holding hands raised up to the sky, and we were bathed in white light. The words come—it is complete.*

The following week this Greenbreath experience continues to live in Rhonda, but now she feels a death, of sorts. After pondering, talking it through, and allowing more healing in, Rhonda realizes this is a death to the pain and suffering that had lingered from her relationship with her husband. Rhonda now understands that she doesn't need physical reconciliation to heal her heart, but that her heart and his heart can heal in the spirit realm with the help of Rose and Tulsi (Sacred Basil). Rhonda says, "I can love him right where I am and this time it can be unconditional." When our hearts heal, not only do our personal relations open up, but the limitless possibilities of the universe also open to us, as this description from a student indicates: "Breathing through my heart a clear infinity symbol comes with my heart at the center, radiating in all directions—up, down, sideways, at angles—always with my heart at the center."

With obstacles released, channels open as your Greenbreath spirals up your spine. Your plant ally spirals kundalini energy up your spine, taking you out of static time and immersing you in the ecstatic experience of deep intimacy with your plant. You experience the bliss of bonding with another being, a green being. Your Greenbreath sets you free. Tina's experience becomes sexual as she says, "Goldenrod came as Ra, the sun god, and presented his golden phallus for me to dance upon. Which I did." At this point you have been breathing for over an hour, and this close encounter with another species is quite profound. It is not unusual to have reached a point where endorphins, oxytocin, and DMT (Dimethyltryptamine) are being released, bringing you to heightened levels of awareness and peak experience.

In the final merging that occurs during Greenbreath, you are free of obstacles, healing wounds of misalignment, and filled with your true essential nature, which is a part of the larger Nature. As a part of Nature you carry the wholeness of Nature within. The Greenbreath of your plant ally has cleared the debris, opened channels to Source energy and healing energy, and brought peace, balance, and harmony. As you resonate in harmony with Nature, there is no more separation between you and your plant, you and Nature, you and Earth, you and the universe. Your plant lives inside you, and you will never be alone again. Taylor says, "Not only am I reborn as a flower, but each of these points of light is also a flower—and also a star—and also an ancestor—and also a photon of communication and entrainment between each of us. And then *I* am each of those points of light, and I am continually shattering into tinier bits of light as my understanding grows. The more I understand, the more I shatter into tinier pieces, until there is nothing left of me in any physical form—there is only awareness and a total merging with it *all*."

You may think that a Greenbreath session is all about your own healing, and of course, it is but it is also much more. Once the plant or tree lives inside you and you are bonded, it becomes clear that your relationship with your plant or tree is reciprocal. You may find that your plant or tree ally calls upon you in some way to do its bidding. Catherine shares her experience of this after a Greenbreath session with Redwood.

> *While driving away a few hours later, I noticed the centers of my palms were burning hot! There was so much energy in them it was startling. Soon I had numerous synchronous experiences on the way home. My favorite was entering the seaside gift shop that I had never been in before and walking right up to a display of odd-looking clear but dark plastic bags containing something indiscernible. I curiously picked one up and read the label—it was a Redwood burl! A whole shelf of them! I had no idea what they even looked like until I read the label. (The burl is now growing abundantly in my kitchen window.)*
>
> *A few days later I was sitting in front of my altar when I closed my eyes and spontaneously began sending a "healing" to someone. The process was quite specific to that person and had to do with guiding them from an inner place of pain to another inner place where they allowed themselves to feel love and thus to experience healing. When that place was reached, I just as spontaneously snapped out of it and opened my eyes as though nothing had happened. It was not "me" doing it; I was simply a messenger. This was a completely new experience for me, and something I had never even thought about. Nevertheless, it continued to happen many times. I felt several friends and family "pulling" on me for help and eventually did*

about seven of these in just a few days, sometimes one right after another. Within a few weeks, I talked to most of them and received a confirmation of their experience. It was both strange and curious. I was like an empty vessel, a container, that enabled Redwood to pour love right through me to others in need. There were some variations of this too, including one that was a message I had to call and relay to the person. Since the Redwood Greenbreath experience, I feel a continuous connection to the spirit world. It is part of my normal orientation now. I can't imagine life without this!

Breath—so simple and yet so profound—without it we cannot live, and when we bring it into our conscious awareness profound healing, alignment, and bonding are available to us, while, at the same time, we are in service to the greater Nature by sharing our carbon dioxide that becomes the green beings' in-breath. As we complete our conscious breathing, we are aware that we are all dreamers, and together we can dream into existence the world we know is possible. So let's dream big!

Head Knowing

The last aspect to engage with in building a co-creative partnership is cognitive, or intellectual, knowledge of your plant ally. Once you have engaged on all the other levels we have previously discussed, then it is time to read everything you can get your hands on and to research the plant's physical properties, history, folklore uses, and magical properties and talk to as many people as you can about their experiences with your plant ally. We save this part of learning about your plant until the end of your investigation because the tendency, in the Western world, is to place more importance on "knowing" the plant through the intellect. My experience of relying on head knowing is that it clouds my perception because most of the information I am taking in is from a source outside myself. I strive to have direct experience and receive direct transmissions. Plants have a direct connection to Source energy, and when I open my channels, I become the "hollow bone"—a Lakota term that means to allow healing energy to flow through one by becoming like a bone that is hollow. Via the plants I, too, have direct access to Source energy. Even though I do rely on my guides, who could be plant spirits, animal spirits, angels, ancestors, or other beings, I am keenly aware that we are in a co-creative partnership, which means I'm an equal partner and get equal say in how to proceed in any given situation. Those who rely too heavily on their "guides" to give them all the answers are essentially outsourcing their decisions instead of tapping into their own inherent knowing, which comes directly from Source.

As we become deeply involved with our green allies, it is so important to view them as the incredible beings that they are, in and of themselves. They are *not* people, but our tendency is to anthropomorphize them. We refer to them as "her" or "him" and give their spirits human form and treat them like our best girlfriend. Is it possible for our interspecies partnership to evolve to the point at which we let plants be their own unique, diverse beings instead of carbon copies of us? On the other hand, Nicholas Epley, a professor of behavioral science at the University of Chicago and world expert in anthropomorphism, is quoted in a Quartz article by Leah Fessler, that "historically, anthropomorphizing has been treated as a sign of childishness or stupidity, but it's actually a natural byproduct of the tendency that makes humans uniquely smart on this planet" and that "anthropomorphism is the byproduct of having an active, intelligent social cognition." Because at our core we are social beings, we seek out others to be with. Sometimes these others are not people but, instead, plants, trees, animals, or particular landscapes. These "others" become our friends, and so we give them human qualities.

Because I'm so closely connected to plants that I consider them my "beloveds" (a human attribution), I am certainly anthropomorphizing these plants, but at the same time I treat them as the unique beings they are. Is it possible to have a loving connection with a plant or animal and still value its unique beingness? I raise this question as something to ponder, in hopes of contributing to a broader understanding of interspecies communication and relationship.

ELEVEN

Facilitation of Plant Initiations

The ceremonial process [initiation] offers a unique way to connect deeply with all aspects of a plant, opening gateways to spiritual realms and facilitating powerful transformation at physical, emotional, mental, and spiritual levels.

CAROLE GUYETT

What Is a Plant Initiation?

I have been working and playing with plants my entire adult life (more than fifty years now) and have experienced much of my bonding with Nature through plants. Because we are so closely related to plants through our breath and as a food source for sustenance and tissue generation, they are naturally an easy avenue to deeply connect with Earth and the other elements. But just like any relationship, we can skim the surface with pleasantries, or we can go beyond the superficial and immerse ourselves in an intimacy that generates an undeniable bond of love, care, and nurturance. I have dedicated my life to being in deep co-creative partnership with plants, while also serving as a spokesperson for them, so I'm always seeking ways to reach further, touch more sincerely, open my heart wider and authentically find common union with the green world. Over and over my experience with plants and trees has led me to believe they are totally capable of leading us to freedom from our destructive ways if we would only take the time to listen and be guided by them.

When working with any particular plant I have encouraged my students to engage with the plant on all levels, which includes ingesting the plant as a food, tea, or medicine (if appropriate) and/or as a flower essence. Another possibility for having frequent contact is to externally apply the essential oil of the plant (if it is safe) or to use it in a spray. I have referred to these practices as plant dieting, meaning to include the plant in your daily diet or routine, but I have come to understand that this level of dieting is rudimentary. In traditional cultures, plant dieting refers to ingesting a plant with psychotropic (mind-altering) properties and usually includes a ritual way of preparing the plant for ingestion. I began to wonder what the possibilities may be for deepening relations with very common plants by "dieting" them in a ceremonial or ritualistic way. I have come to refer to this ceremonial dieting as plant initiations.

Initiation is an ancient form of ceremony that brings one into maturity and results in taking up one's place within the collective. In traditional society, initiation was a coming of age ceremony where the young person became an adult, stepping into their role within the "tribe." The elders would perform the initiations, which sometimes took up to a year to complete and always culminated in a great ceremony and then a feast. In modern times the initiation process has been mostly forgotten, so most adults alive today have never had the opportunity to come into their maturity, bringing forth their gifts in order to serve the community. The elders, too, are mostly gone, so even for those who would like to be initiated into being truly human the opportunity does not exist. In my deep work with plants I continue to be amazed by how they step up to the plate and provide us with exactly what we need. So what is emerging is that the common plants are taking up the role of the elder and offering us initiation into what it means to be truly human, living sanely within the collective. A plant initiation is an opportunity to work deeply with a plant or tree, where we not only receive personal healing but also become bonded with the particular plant and receive the plant's healing gifts, which we can then apply in our work with ourselves, others, or the planet. This form of intimate sharing with a plant allows for us to receive guidance on many issues, open doors of perception, undergo healing on all levels, and expand our consciousness.

Primrose Introduces Me to Initiation

The story of how I came to experience what I'm now referring to as plant initiation may seem a circuitous path but often, when we pay attention, we realize there is a vast web of interconnection that leads us to exactly where our soul is destined to travel. Several years ago, in the early spring, I traveled to Ireland and England to teach and to spend time with my dear friend Carole Guyett. The first leg of my journey was to County Clare, Ireland, where I had the great fortune to participate in a sound healing with a group Carole has been working with for years, called the Labyrinth Healing Group. This group of people have been dreaming, journeying, doing ceremony, and working with the deities of the labyrinth for years. Their work has evolved to include the plants that grow in the labyrinth and the songs that emerge as tonal sound, flowing through them for healing purposes. During this day of sound healing, I was both a participant receiving a healing and one of the persons allowing the healing sounds to flow through me to others who had come for healing. Many levels of healing took place for me that day, but what stands out is that I was being prepared for another aspect of my work with the plants and their magnanimous spirits, even though it was not exactly clear then what that was.

After my time in Ireland I traveled to England to teach two classes, one on Greenbreath and another on plant communication. While in Cornwall, I stayed in a lovely little cottage nestled in moss-covered trees beside a meandering brook next to a hillside covered in the most beautiful pale yellow flowers. I was immediately smitten by these little plants with their heart-shaped petals comprising their yellow flowers. A couple days later, I was in the Midlands teaching a three-day plant communication course, and while class participants were out communing with their plants, I wandered the grounds and gardens and there again was my little yellow-flowered plant. I took this opportunity to begin building a relationship with this plant that I discovered to be Primrose. While I was swooning over Primrose, powerful forces were at work. As I related in chapter 5, during this week a phenomenal occurrence came in the form of a volcanic eruption in Iceland that generated such enormous amounts of volcanic ash that airplanes were being advised not to attempt to fly through it. By Sunday afternoon, at the close of the workshop, my reality shifted from that of falling in love with Primrose to *I can't get home*—slam—what a shock to the system. After spending a harrowing twenty-four hours in London trying to figure out where to stay and how to get home, something magical happened. The prospect of staying in London was dismal at best, so I opened my heart to the place where all possibilities exist, and there was a heart-shaped yellow glow. I was able to contact Carole in Ireland, and she told me how to take a ferry to get back to Ireland. After three trains and an all-night ferry across the Irish Sea, I arrived in Ireland. I had found my way home—home to the land of my ancestors and my heart—Ireland.

Now, you must be wondering what all this has to do with plant initiation. The truth is: life is really about the journey, not the arriving. This journey took me back to Ireland to be with Carole for another four glorious days where I learned about a deeper form of plant dieting. I already knew that Carole was working with plants ceremonially and that she was making certain preparations for ingestion in order to broaden the potential of experience available through the plants. Essentially, she was experimenting with working with common plants to alter one's field of perception and to allow for an expanded state of awareness. Here I was back in Ireland, with a seemingly magic thread being woven into the fabric of my life that was totally unexpected.

Carole was preparing for a plant dieting with her Labyrinth Healing Group, which was to take place on Beltane. Of course, the plant they were dieting was Primrose. During this time together, I helped Carole gather the Primrose blossoms that were to go into the dieting elixir, and she sent me home with a bottle of the elixir that she had prepared in a special way, along with instructions for ceremonially dieting the plant on Beltane, the same time her group would be working with Primrose. Luckily, I had time to spend with Primrose while at Carole's. How fortunate this was because I do not have it growing at home. My experience of Primrose in person is that she has a subtle, sweet fragrance with pale yellow flowers, with five heart-shaped petals and a darker yellow sunburst radiating from the middle. The leaf is bumpy, like bubble wrap on top, and very hairy on the backside. It's very soft with such delicateness and sweetness, making me take a deep breath, sigh, and relax.

I had a dream prior to the diet where I could see all the auras of the trees, and they were like swirling clouds. I wondered if the impending Primrose diet had anything to do with this dream. On the day of the diet, back home in Vermont, I made an altar and put the dried Primrose flower and leaf I had brought from Ireland in the center. In preparation, I called in the directions and all my helpers and guides to be with me. I drew a personal tarot card, which was the Nine of Wands, and a group card (I was connected during this diet with the Labyrinth group in Ireland), which was Temperance. Both cards have to do with being fully in one's creative power.

I have come to call this a plant initiation, instead of a diet, as it really *is* an initiation. As Carole describes, "It marks the crossing of a threshold and signifies movement to the next level." That's what I see this form of plant dieting doing—taking us to the next level of relationship with the plant, with ourselves, and with spirit.

This Primrose initiation begins by me toning, with each of my chakras coming solidly into my heart space. I drink the first draft of elixir and sit with the fire, where there is an amazing green flame touching deeply into my heart. I've never seen a green flame quite like this come from a fire. I drummed and sang and jumped the Beltane fire, leaving behind lethargy and that which keeps me from being all I can be. Instead, I jumped into "being life."

Dreamtime with Fairy Queen Primrose

I walk my labyrinth at the same time the folks in Ireland are walking theirs, receiving answers to the questions that were being asked: Who am I? "I'm Shares the Flower Song." What is your intention for being here? "I come to deepen my ways with plants and to recommit through deep ritual. I come to learn and receive the gifts of Primrose." What are you creating and manifesting in your life? "I'm creating a healing sanctuary for people, Earth, the plants, the elements, the directions, the energy wheels, points and lines within and without. My manifestation is a sanctuary here and a book to tell others how to create their own or to go to established sanctuaries to bring back balance." What is your connection to Primrose? Who is Primrose? What does she bring and what does she ask for? "I'm only beginning to learn who Primrose is. When in England and Ireland I became somewhat acquainted and was able to be with the live plant. I plan to plant Primulus vulgaris *here. Primrose seems to hold the key to unlock the door to inner riches and treasures, one's true essential nature. She asks for us to wake up—wake up and open to these gifts." What does your heart tell you? "It tells me not to get distracted. Do what spirit says—to speak through the heart." What is your power? "My power is my love—my love for the plants, Earth, the elements, the land, spirit." Are you willing to transform? "Yes!" How? "I will walk my talk. Practice what I preach. Recommit to plants by working with them ceremonially." As I complete the seven circuits of the labyrinth, I drink of Primrose again, feeling her beginning to work with me because I have agreed to transform.*

In my dreams this night Primrose has made a bridge between Sweetwater (my home) and Derrynagittah (Carole's home). There is a key ingredient I'm not supposed to miss, but as the dream wanes, I cannot grasp what this key ingredient is.

In the morning, I know everyone in Ireland is sitting with Primrose outdoors. Even though I don't have the Primrose plant to sit with, I take my drink of Primrose elixir and return to Primrose in Ireland via the daydream, and I see her beautiful heart-shaped yellow petals and the bumps on her leaves that are multifaceted like a crystal ball. I get a hot, heady sensation with a band of pressure around my eyes and ears. The front of my ears and jaw are tight as a drum. Drumming begins slowly across this band—a constant, slow beat, not fast. The drumming sound is a key, and then I hear the symphony of the stream flowing near where I sit. Light is another key—light shining through a crystal ball in my garden is refracting into a prism, which seems to reflect the prism of light of the leaves of Primrose that I see in my daydream. My vision becomes blurred as my eyes keep losing focus—I'm "seeing" instead of looking. Primrose is a very powerful, radiant being. I'm almost afraid of its radiance. It shows me the possible and potential radiance that I carry and can shine forth. What a huge responsibility to be true to your radiance! Primrose shows me the importance of walking my soul's path and not getting distracted from it.

Mark and I go to the forest, and he rattles for me near the magic waterfall and rock formation so that I can go on a shamanic journey. Lela (my wise woman within) takes me to a tiny cottage. I ask her how I am to get in. She says you have to make yourself small. I ask her how to do that, and she says to listen to the rattle. I become small, like an insect, so I can fit into this very tiny cottage. There is an old woman inside, and I ask if she is the spirit of Primrose. "Lord no!" she says. "I'm to give you a cup of Primrose tea." She gets out these tiny cups and gives me tea. When I finish, she takes a big key off the wall and tells me to follow her. I follow her out to her garden where there is a big wooden door rounded at the top. At some point the sounds around me have changed, and I ask her about this. She says that I am now in the land of the fairies and that many things are different here. She says, "You do know Primrose is of the fairies." Then she unlocks the door. When the door opens a dazzling light comes through. I go inside, and there are staffs and wands with crystals in them all over the place and piles of gold everywhere. A beautiful fairy queen appears and asks if I would like one of the wands. She picks out a beautiful wand or staff with a clear quartz crystal point on the end and embedded jewels and stones on the shaft and hands it to me. There is gold spiraling around it. It is so beautiful. I ask if I need a wand or staff in physical form, and she says it may come to me someday. As I thank her and am about to leave, she says, "Now you must serve me well." I agree and go out the door, back to the cottage, and then back with Lela to return to present reality. What a beautiful journey, and I have been given a gift from Primrose.

My Greenbreath session with Primrose is not terribly eventful. Some fear arises that I won't be able to live up to Primrose's expectations of me. During the remembrance song Primrose, as the fairy queen, leads me to the land of the Sidhe, the land of fairy folk of Ireland who live in the mounds. I don't go in—would I have come out? I feel good in my heart during Greenbreath as Primrose dances with me and lies beside me. The elixir definitely makes me spacey.

We continue our dreamtime work with Primrose with a dismemberment journey. Carole says of this journey, "Dismemberment can be a common experience when working deeply with plant spirits. It brings healing and is a way of dying to the ego in order to become more one's true self." During the journey I go to the garden again, where the little old lady unlocks the door, and it is dark inside except for one candle. I can't see Primrose, but she tells me to get on the table. I tell her that I'm afraid. She doesn't understand why because I will get put back together even better than I was before. She says she will cut out the lazy part that leads to not being authentic, the part that needs to take time out to be with plants and spirit, instead of it being a part of my natural flow. Tulsi, Artemisia, and Hawthorn come to be with me during the dismemberment surgery. It doesn't really hurt. Once Primrose has taken me apart, these supporting allies who have shown up to be with me take all my parts in a basket to the stream here at Sweetwater to make sure I am put back together properly. Tulsi, my soul, Hawthorn, my heart, and Artemisia, my energy body, make sure I am put back together. Once I am back together again, I find that I am in the land of Primrose, the fairy queen. She is her beautiful self again. She gives me a wand identical to the one she had given me in yesterday's journey. She says this one has the power in it, and immediately the pad I am lying on literally bounces up and down. She says she knew I wasn't ready for the power-filled one before, but now I am ready.

It is morning again, and I am drinking my final draft of Primrose elixir, for a total of eight ounces. The final shamanic journey is to the temple of Primrose. I follow a yellow brick road, accompanied by Tulsi, Artemisia, and Hawthorn, to the Emerald City castle. Fairy Queen Primrose greets us. Everything has twinkles of light on it. She takes me to a solarium that again is light filled, with crystalline twinkling lights. She tells me she can make a bridge to any land, just the way she has made a bridge between Sweetwater and Derrynagittah. I ask, even if she doesn't grow wild in the particular land where I live, can I still visit with her? She says, this is dreamtime space and she is the spirit of Primrose and her gifts are outside time and space. She also tells me that I need to give equal accord to the dreamtime space as I do to ordinary reality as so much can happen in the dreamtime. Something can't manifest until you dream it. She reminds me that my power is my love.

This time with Primrose is very powerful, and I feel that drinking the elixir adds an altered dimension to the whole experience, by allowing easy access to the spirit realm of Primrose. Even though much of the time is spent in the dreamtime dimension, the quality of the encounters with Primrose are quite vivid and seem just as real as waking reality. Her messages are direct and the images clear. I feel I have gained not only an ally, but a mentor whom I can call upon at any time for help and guidance.

Preparing for a Plant Initiation

Since my first initiation experience with Primrose years ago, I have now facilitated several plant initiations, both at my Sweetwater home and in other locations around the globe. The initiation ceremonies I facilitate are over a three-day period, during which we experience the plant or tree in many different ways. We begin by preparing ourselves ahead of time by eating certain foods and refraining from certain activities. Entering an initiation with a clear body, mind, and spirit is a way to honor the plant elder, who is the teacher and initiator. It is with utmost respect that we approach these magnanimous beings as they lay the fertile ground for a deep transformative experience.

Once we enter into ceremonial space by setting the container, we begin with ritually mixing our elixir of different plant parts and mediums (primarily water and alcohol) that they are soaked in. Each participant in the ceremony helps to stir the elixir, which is in the biodynamic style of stirring in one direction and then switching to stirring in the other direction. This method of stirring creates a vortex where both heavenly and earthly energies can be incorporated. This elixir will be ingested in the amount of two ounces taken eight times during the initiation to enhance the consciousness-expanding qualities of the plant.

We deepen with the plant in many ways, including dreaming, journeying, and taking on the essence of the plant, aided by masks we make and then wear during a fire council where we embody the plant. We also experience the plant through Greenbreath and stepping into the initiatory experience of dismemberment, which is a traditional shamanic way of dying to the old and being reborn into a new you. Our plant ally will aid us in ways known and unknown as we leave space for the magic and mystery to emerge.

Plant initiations are probably *the most* powerful way to build your relationship with plants and trees into a co-creative partnership where the plant or tree becomes your beloved for life and one who supports you to be your authentic self, journeying on the path you came to walk this time around.

Preparing the Elixir

The preparation of the elixir could begin an entire year prior to the initiation, assuming you are harvesting the plant yourself and making an alcohol-based tincture or drying the plant to make a water-based infusion. The point is that you have to plan ahead. Even if you don't have access to the plant to harvest, you will still need to acquire the plant materials from someone. So for example, if you are including tincture in your elixir, then ideally you tincture it fresh, which means finding a reputable company you can buy it from who will ship it to you immediately after it is harvested. Some roots and berries can be tinctured when they are dry, but fresh is best. For making infusions for the elixir, dried plant material is preferable, and you can get this from a reputable herb company relatively easily. Common mediums are water (infusion, hydrosol, essence) and alcohol (tincture in vodka or brandy). Red shiso vinegar can be used to make a tincture if you absolutely don't want any alcohol. Sometimes you might make a syrup- or honey-infused preparation. You can also incorporate an essential oil of the plant (when appropriate) to bring in the aromatic quality of the plant, which you can put on bits of cloth for participants to smell.

I like to work with as many different plant parts as possible—flower, leaf, root, and berry (when appropriate). Sometimes bark may be included, especially if it is a tree. Begin gathering these parts as early as you can. Usually, there is some kind of complementary fruit juice included in the elixir as well.

Note: Proceed with reasonable caution when ingesting plant elixirs and do research as applicable before commencing a plant diet. Please be aware that doses during plant initiations are small and introduced incrementally. If you are concerned, after researching the plant that is serving as the initiator, about ingesting the elixir as described here then you can also work with the essence of the plant. The essence of the plant is the flower, a little bit of leaf, and potentially other parts of the plant floating on top of spring water and sitting in the sun for two hours. Later in this chapter details for making a plant essence are given. There are no chemical constituents in this essence thus no chemical reactions in one's body. You are only receiving the vibratory resonance of the plant, which is just as potent as ingesting the elixir. Please consult with your health care provider as needed before commencing a plant diet.

During the initiations I facilitate, each participant ingests 16 ounces of elixir over a period of two and a half days. So, if you have twelve people doing a plant initiation together, you will need a total of 192 ounces. I like to have some extra, so I would increase this amount by 8 ounces to a total of 200 ounces (one and a half gallons plus an extra 8 ounces). An example elixir with a total amount of 200 ounces might contain:

16 ounces tincture
128 ounces infusion
8 ounces syrup
16 ounces hydrosol
32 ounces fruit juice

The tincture and infusion can be a mix of the different parts of the plant, for example, 64 ounces of flower infusion and 64 ounces of leaf infusion, making a total of 128 ounces of infusion.

The composition of the elixir is determined by which plant or tree you are working with, the plant material you are able to acquire or what you have yourself, the time of year the initiation is taking place, *and* what the plant or tree is guiding you to do. Remember, the plant is the elder who is leading the initiation, so you need to pay attention to how the elder wants the initiation to go, which includes what goes in the elixir.

I like to let the elixir sit for a couple hours in the middle of my labyrinth, which serves as a living temple, so that it becomes infused with spiritual energy. If you don't have access to a labyrinth, then you can create a spiral shape (a long string or rope work well for this) and place the elixir in the middle. Both the labyrinth and spiral are geometric patterns that amplify life force.

Setting the Container; Mixing and Ingesting the Elixir

Now we are ready to begin the initiation by setting the container, which is a way to create a safe and sacred space within which we can be initiated by the plant elder. Prior to participants arriving, I have already created an altar on a cloth at the center of the circle where folks will sit that includes the four cardinal directions and the elements represented along with a vase of flowers in the middle and cornmeal and tobacco to make our prayers. Each participant (or just myself if this is a solo initiation) is cleansed with the smoke of Mugwort, and the spirit of each direction—east, south, west, north, below, above, and center—is called into the circle. Then each participant writes down their intention for the initiation and puts it under the altar cloth and places an item on the altar that they have brought with them for this purpose. This item is something that might come from their own altar at home or it carries some type of special significance. When an item is placed on the altar, it absorbs the energy that is created during the ceremony. The final piece of setting the container is to ask for another plant besides the initiator to give us support during the ceremony. We draw a card from a deck to discover who our support plant is.

Now it is time to mix our elixir in a large pot with a long spoon that can reach to the bottom. The stirring method we use is akin to a biodynamic gardening stir, which creates a vortex spin, bringing in both the cosmic forces and the earthly forces. We stir in one direction and switch to stir in the other direction while speaking our prayers into the elixir. One person begins stirring, and the rest of the initiates drum, rattle, or play some type of instrument to raise the energy and activate the elixir. Each person comes to the pot, and when a person is done stirring, they give a nod to the next person in the circle to come forward and begin stirring, never letting the vortex stop spinning. When we come to the last person, the music becomes louder, and we bring the energy up to a crescendo until we stop. We circle closely around the pot and give our breath to the elixir. Now our elixir is ready to be ingested.

Using a beautiful decanter, we pour enough elixir into it for everyone to have one 2-ounce drink. The elixir is served in a ceremonial fashion, with each person approaching the facilitator with their glass and receiving the elixir as a sacred sacrament. Once everyone has received the elixir, we take the first sip of the elixir together, paying close attention to what is happening on a physical, emotional, mental, and spiritual level. We let the elixir course through our body, filling us with its essence. We will drink in this manner a total of eight times throughout the initiation, each time paying attention to how the plant is working with us.

Deepening with the Plant

I have discussed deepening with a plant by communicating via sensory awareness, felt sensation, and being in the dreamtime with the plant, as well as the Greenbreath journey. (I also extensively discuss deepening with a plant in my book *Plant Spirit Healing,* pages 82–102.) So I will not share this part of the initiation process even though we do engage in these methods with the initiating plant. I will add to the ways we deepen with the plant by discussing parts of the initiation process that have not been written about elsewhere.

Embodying the Plant

One of the ways of getting to know an aspect of Nature is to embody that aspect. In order to do this during an initiation, we make masks that portray the unique gifts of the plant that we are called to bring forward. We generally make masks in the afternoon of the second day, so these gifts come from the experiences we have already had with the plant to this point. We decorate our masks with paint, fabric, natural material like feathers—really, we can incorporate anything we feel called to use as a material. All of the masks from the initiates are so different and incredibly beautiful, and by the time everyone is done, we are a community of unique plant beings (see color plate 8). We then wear our masks during a plant council, which is held around an evening fire circle. This type of council originated from the Council of All Beings work brought forth by John Seed, Joanna Macy, Pat Fleming, and Arne Naess. In their book *Thinking Like a Mountain: Towards a Council of All Beings,* the council is described as "a form of group work which prepares and allows people to 'hear within themselves the sounds of the earth crying,' a phrase borrowed from Vietnamese Zen master Thich Nhat Hanh, and to let other life forms speak through them. It is a form which permits us to experience consciously both the pain and the power of our interconnectedness with all life."

In preparation for the council, we don our masks, dress in ceremonial clothes, and stand in a circle to witness and be witnessed. Each person is seen by the other "plants" and also sees all others. In this way, we acknowledge that we, as representatives of the plant elder who is initiating us, are totally seen and honored. This exercise, called "I See You," touches one's heart because being seen for who one truly is can be one of the deepest communions one experiences.

We attend the council as plant beings and each "plant" has the opportunity to speak, elucidating their concerns, inspirations, and suggestions for moving forward in our plant work to shift the imbalance on the planet, bring about thriving for all beings, and help humans take up their rightful place as a part of Nature. Every "plant" voice is heard in council, and they are never referred to as humans because in this milieu they are plant beings, speaking as and for the plants. Reference to humans are made only in the third person. Once the council is completed, we transform back into humans and carry with us the many perspectives presented in council. Though this may seem like a form of playacting, it is a profound way to experience the philosophy of *deep ecology*, a term coined by Arne Naess, which advocates that we move away from anthropocentrism and toward a credence that humans are equal to Nature, not higher or better than, and *all beings,* both human and nonhuman, have the right to thrive.

Walking the Labyrinth

A classical labyrinth is an ancient geometric pattern dating as far back as 2000 BCE. Labyrinths can be found in Chartres Cathedral in France, in China, in the Southwest of the United States, and throughout the British Isles, as well as many other locations across the globe. Traditionally, walking a labyrinth is used as a meditation, which weaves left to right and back again while walking to the center and then walking out again. You can choose to walk the labyrinth when you are seeking a solution to something that is troubling you, or you may want to commune with your wise person within or with a higher power other than yourself. My experience of walking the labyrinth is that when I finish walking to the center and then back out, I feel refreshed, have more clarity, and am at peace. I see the labyrinth as a living temple where I have direct access to spirit or the divine.

During plant initiations one of the activities we do is walking the labyrinth, with questions asked of initiates at the entrance, midway point, and in the center. The questions are not to be answered immediately but to be contemplated to see what may bubble up. The question asked at the entrance is, "Who are you and what is your power?" The question asked midway usually has to do with the particular attributes of the initiating plant. So, for example, the question asked during a Hawthorn initiation may be, "How will Hawthorn help heal your broken heart?" The question asked at the center is usually, "Are you ready to transform your life and, if so, how will you do this?" The contemplation of these questions offers an opportunity to look deeply within and perhaps even shine light on shadowy parts of yourself.

If you don't have a labyrinth to walk, you can use a rope to create a spiral path to walk. I like to energetically activate the spiral by placing crystals in the four cardinal directions. There is no need to activate a labyrinth since its geometric pattern serves this purpose (see color plate 9).

Prayer

Sitting alone, initiates create prayer arrows from a stick that presents itself to them and yarn to wrap the stick. Each time we make a wrap of yarn around the stick a prayer is spoken from our heart. We pray that we may be worthy of the gifts bestowed on us from the plant elder. We pray about all the times we have been worthy and all the times where we haven't. We ask for forgiveness for all the wrong we've done in the world and we forgive all the wrong that has been done to us. Making prayers with each wrap, we are in deep gratitude for all the good we have done in the world and all the good that has been done to and for us. Our prayers are multitudinous, microscopically honest, and come from our authentic heart.

Singing

Because sound is one of the foundational ways in which communication takes place, some type of engagement with sound is crucial. Even though it may seem like the sound is coming from us, it is a translation of phonons coming from the plant. When we create sound together, we can align our vibratory resonances and ride the same wavelength. In small groups we share all the gifts we've discovered about the plant elder and work together to create a song to honor the initiating plant. Or maybe we are alone and we exchange tonal variations with the plant, until the tones flow like water over stone creating a beautiful natural melody. The other possibility is to listen to the plant's song with a Music of the Plants device (available through the website Music of the Plants) that, via electrodes hooked up to the plant, reads electrical impulses, translates them into musical notes, and plays the vibratory resonance of the plant through speakers.

Dreams

Before going to bed at night we ask the plant elder to bring us a dream. This does not necessarily mean we dream about *the* plant. It simply is that the plant brings a dream that is significant for us at this time. The key is that we remember the dream when we wake up in the morning and then strive to derive meaning by paying close attention to our intuition and synchronicities that may offer helpful interpretation. Sometimes we don't remember our dreams so we pay attention to the feeling sense that we wake up with. The dream the plant elder brings during an initiation is usually quite significant and can help us unravel otherwise confusing bits of our experience.

Dismemberment Journey

Another way that we ask for the help of our plant elder is through what is called a shamanic dismemberment journey. With the aid of a rattle, initiates are guided to a place in the woods where they meet a wise old woman who is the gatekeeper of the realm of "dying to the old and being reborn to the new." She takes the initiate to the gate and leaves them there, where the plant elder awaits behind the gate. Slowly the gate opens to reveal magnificent light pouring out of this realm of the beyond. The shape of the plant elder emerges, and she or he guides the initiate to their healing chamber where the initiate will be transformed. During this journey the plant elder takes the initiate apart and then puts them back together again. What no longer serves the initiate on their path is removed and what does serve the initiate is woven into their beingness. When the initiate returns from the dismemberment journey, they are a whole new person. At this point in the initiation the initiate has been given many gifts by the plant elder and is ready to make big changes.

Reciprocity

We are now at the end of our initiation, and it is time to give back to the plant or tree we have been engaged with. So many gifts have been given to us, and we are grateful beyond measure. Now, we reciprocate by giving a gift that we have made or a gift that holds a special place in our heart. We circle around the plant or tree, and each initiate approaches and speaks eloquently to the plant, which has become their ally. We speak from our heart, showering our plant ally with love, honor, and gratitude for all that they are in the world and all that they have given to us. We leave our gift with the plant or tree, which serves as a confirmation for all who pass by that this is a plant elder that has stepped forward to initiate us into being truly human. We have moved one more step further in our spiritual awakening because this plant elder saw our need and called us to be initiated (see color plate 10).

Plant Essence

To continue to support the healing process that began during the initiation, each initiate is given a dropper bottle containing the essence of the plant to take home. The instructions are to take the essence for at least six weeks. The average dosage is four drops under the tongue three times a day, but this could vary depending on what is needed. I suggest that you carry the bottle of essence in your pocket and take the essence as often as necessary. If you are doing a solo plant initiation, then you may want to make your own essence instead of buying it from a store or flower essence company. The difference between a plant essence and a flower essence is that other parts of the plant are incorporated into the essence, not just the flower. So, for example, I will include a little flower, leaf, stalk, or berry (if these are present on the plant or tree) to be a part of the essence.

The following are the steps to make the essence:

Fill a small, clear glass bowl with spring water and, asking permission from the plant, carefully add each of the plant parts so that they float on the surface of the water.

Walk a labyrinth or spiral circuit with the glass bowl, receiving communications from the plant spirit as you walk and asking the spirit of the plant be infused in the essence.

Set the glass bowl in the center of the labyrinth or spiral circuit for two hours if there is bright sunlight, or four hours if it is cloudy.

When finished, retrieve the essence by walking out of the labyrinth or spiral circuit, again receiving any 'downloads' from the plant.

To make the essence into a preparation:

Pour off the water from the glass bowl. This is called potentized water.

Fill a bottle with half potentized water and half brandy. You can substitute brandy with red shiso vinegar (you will find this in an Asian food store). This preparation is called the mother essence.

Fill a 1-ounce dropper bottle with half spring water and half brandy and three drops from the mother essence. This preparation is called a stock essence bottle.

Fill another 1-ounce bottle with half spring water and half brandy and three drops from the stock essence. This preparation is called the dosage essence bottle.

The dosage essence is handed out to initiates at the end of an initiation and it is the bottle from which you take your daily drops that go under your tongue.

Closing the Circle

Now it is time to close our ceremonial circle. We do this by passing a talking stick (which could be an actual stick used ritually for this purpose, or another object) from person to person. When you are the person holding the talking stick, you speak from your heart while everyone else witnesses you. There is never cross talk during the passing of a talking stick. Each person asks this question of the next person as the stick passes from person to person, "Sister (or brother), how will you bring the gifts of (insert plant or tree name here) into the wider world?" After a person speaks, together we say out loud, "So be it."

Now that the sharing with the talking stick is completed, we end our ceremony by singing a song, which also serves to thank the spirits of each of the directions. The words of the song are, "May the circle be open but unbroken, may the love of the Goddess be ever in your heart. Merry meet and merry part and merry meet again." This is repeated three times.

What I have outlined here are suggestions for your plant initiation. I allow myself to be guided by the plant elder as to what activities we engage in during the initiation. During the Hawthorn initiation I was instructed to bring the initiates on a "pilgrimage" to the Hawthorn tree across the road. Other plants may ask for a plant *limpia* (a Spanish word meaning "clean"), which is a type of spiritual bathing. To make a limpia, we offer specific prayers to the initiating plant, then the plant is combined with water, which is then sprinkled over and around our body, or a bundle made from the plant is dipped in water and then sprinkled over the person. There are many other types of activities that the plant or tree may want you to engage in, so you need to pay attention and stay flexible, remembering that the plant is performing the initiation and you are either the facilitator or the initiate—you are not in control. The plant or tree has called the initiates to participate and knows what they need at this time—our role is to trust this.

More information about plant initiation facilitators and the initiations they conduct can be found at my website wakeuptonature.com. Other resources for plant initiations are: Carole Guyett at derrynagittah.ie located in Ireland and Emma Fitchett (Farrell) at naturalesoterics.org located in the UK.

PART FOUR

Plant Initiations

Green, how I want you green.

FEDERICO GARCÍA LORCA

TWELVE
Hawthorn

Crataegus spp.

When I begin to contemplate the next initiation that I will facilitate, I pay attention to which plant or tree speaks up and lets me know that it wants to be the initiator. This occurred loudly and clearly when Hawthorn stepped forward during the summer prior to the following spring's initiation. I was sitting on the other side of the labyrinth when my gaze fell upon the young Hawthorn tree that had been given to me by a group of students at least ten years ago, if not more. As I entered into the daydream of Hawthorn, I mused, "You know, I have always wanted to do a Hawthorn initiation with you here at Sweetwater, but you never flower." As previously mentioned, Hawthorn sent its touches to me and I interpreted, "Well, if you would facilitate an initiation I might flower." I took this in and had a bit of an aha moment. I replied, "So does this mean you want to be the initiator next year?" A simple yes was the reply. From that moment, the Hawthorn initiation was set into motion within the field of dreams.

It so happens that most of my interaction with Hawthorn, up to this point, was with a large Hawthorn tree that grows across the road from

Sweetwater Sanctuary. This is the tree I harvest berries from each year to make into tincture and to dry. When it became clear that Hawthorn was to be the initiator the following year, I was thrilled that, when I went to harvest berries, I could commune with this Hawthorn about the upcoming initiation. As I was harvesting berries and chatting away with Hawthorn, what I received was, "Yes, it will be auspicious when you come to me on a pilgrimage." I stopped in my tracks and repeated back, "Pilgrimage?" I was a bit stunned and said, "So what would that look like?" Hawthorn responded, "Well, it would look like: you and all the initiates will walk from Sweetwater to here on a pilgrimage." I came away from that encounter amazed at this instruction from Hawthorn. I have learned over the years that, when a plant or tree gives very clear instruction or guidance, it is important to not only listen but to take action if it is needed. In this case, it was a very big action of going on a pilgrimage.

Because I was informed by Mama Hawthorn across the road that we were to come to her on a pilgrimage, during my preparations much of my attention was focused on her. But as the initiation drew near, I realized I needed to spend time with the younger Hawthorn here at Sweetwater. As I approached her, I glimpsed something in her branches on the far side, and as I came full circle around her, I gasped. Here, in the springtime of the year, buds were forming. I exclaimed, "Wow, you are going to flower!" She immediately responded, "Wasn't that our agreement—that I would flower and you would facilitate an initiation where I am the initiator?" I said, "Yes, but you've never flowered before and yet here you are, about to flower." I couldn't believe my eyes, and one more time, I was astounded at these amazing sentient beings. This was co-creative partnership at its best—together we agreed to participate in making this initiation happen in a powerful and heart-felt way. This was one of the profound gifts that Hawthorn gave to me—I experienced deep insight, a new level of trust and a communion like no other. I began to weep as I realized deep in my bones that I can trust my plant and tree kin completely. I don't need to carry doubt anymore as this knowing now becomes inherent.

Pilgrimage

Usually, during our initiations, we spend time with the plant or tree, engaging our sensory awareness and felt sensation while being in the daydream. During this initiation, we walked from Sweetwater to Mama Hawthorn on pilgrimage and spent our sharing time there, under the branches of this majestic tree. Pilgrimage, in the traditional sense, is a spiritual journey, usually to a particular place and, oftentimes, is undertaken alone. Interestingly, I got a splinter in my big toe just before leaving to walk to Hawthorn, so I was a bit compromised on my pilgrimage. This helped me to understand that "the journey" can be painful and yet one must continue on. Once I was sitting with Mama Hawthorn, I observed that one of her most pronounced aspects is her thorns, which are prolific and very sharp. I immediately saw the solution for my painful toe—I will remove the splinter with a thorn from Mama Hawthorn. I began to dig out the splinter and successfully removed it. I then remembered a teaching from my teacher Rocío Alarcón—to manage the forces I must focus on the solution, not the problem.

As I sat under Mama Hawthorn I asked why she suggested we come on a pilgrimage to her. She replied, "Oh, this is not about me, this is about life. I wanted all the initiates to begin to understand that life is a pilgrimage. Your life is a spiritual journey." With this, I began to go through a review of my life, remembering all the firefly moments—the times when a sparkling light appeared, leading the way through the darkness and guiding me to take the path that would be in service to my Earth walk this time around. This morning, in the hour before dawn, I stepped out onto the porch and was delighted by the thousands of fireflies in the field. What an awesome sight! I remembered Hawthorn's words from the pilgrimage and that I'm still on my spiritual journey through life, which seems to be getting more pronounced the older I get. As I watched the early morning light show I was reminded that all I have to do is follow the light to continue to be on a lifetime pilgrimage.

Individuality within Unity

As we move into an era of unity consciousness, Hawthorn reminds me not to lose sight of diversity and uniqueness. In looking out for the well-being of the tribe, don't forget the individual who needs to be seen. This importance of individuality became apparent during the Hawthorn initiation because there were two distinctly different Hawthorn trees to engage with. Mama Hawthorn across the road was old and big and asked for us to come on pilgrimage. The other tree, growing at the edge of the labyrinth, was young and just coming into her flowering and wanted us to know she was a rising star, as indicated by the name she declared for herself, Estrella (Spanish for "star"). Both trees carry the gifts of Hawthorn on a species level, and yet at the same time each tree embodies individual qualities and unique gifts.

We all come into this Earth walk with inherent gifts, and it is part of our evolution to share these gifts with people, animals, plants, and all of Nature. These gifts are part of our true essential nature, and we live from the truth of our being through their expression. Bringing these gifts forward into unity consciousness, we embrace our unique individuality as we step into the Great Oneness. The spirit of Hawthorn said, "I see you," and shared this poem with me:

Like the mist rising on an early morn
I wake up to greet your warmth and light.
Like the spider's web with dewdrop jewels
I see how you make relations.
Like the light dancing on water
I see how you sparkle.
Like the rhythm of the rain
Your heart beats strong.
Like the fragrance of the Lily
Your sweetness is a delight.
Like the full moon rising
Your radiance is breathtaking.

What an incredible tribute to my gifts!

Nonjudgment

A predominant attitude in this modern-day society is that of "us and them," which automatically sets up and maintains a paradigm of separation. When I find myself in this frame of mind, I'm usually being judgmental, which takes me out of heart coherence. I lose my center and get out of balance. Once I've lost my balance, a negative rap begins in my head, and a type of internal war is set into motion. This war is actually a fantasy, but one that plays out with characters and dialogue and takes over my attention. What I have found is that to be in us-and-them mode sucks my valuable energy, which, the older I get, has become very precious to me. What a ridiculous way to spend my time—fantasizing about a scenario that maintains the illusion of separation between me and other people and from the vital life force. This heartbreaking internal world of conflict is antithetical to my desire to live in unity consciousness. It also robs me of the peace that allows me to move with grace.

I asked Hawthorn how I might shift away from internal warring, whether it is with myself, others, the government, the climate, or any of the other myriad possibilities. Her response was to learn to speak with love language. Language is powerful, and when we use warring or violent language, we are perpetuating this negative energy, but when we engage in love language, we are disseminating benevolent and loving energy. As I go through my day and the old habit of using cliché phrases arises, like "killing two birds with one stone," "ride shotgun," "take a stab at it," "choose your battles," or "call the shots," I consciously raise my hand as if it were a stop sign and, in the moment, rephrase what I'm saying to shift from militaristic or violent language to life-giving language that makes me feel good instead of diminished. The more I am consciously aware of my language and shift my words to reflect what I most want to create, I can move into what Charles Eisenstein refers to as "the more beautiful world my heart knows is possible." In this space I can "be love," where there is no separation from myself, others, Nature—all of life.

Forgiveness

Hawthorn brings forward our ability to give positive impulses to the heart, which supports the heart's shift into a state of coherence, where the gut and brain become synchronized. Forgiveness is a powerful spiritual practice that ultimately leads one to experience compassion.

During a recent Hawthorn initiation, Shara describes her experience.

I asked Hawthorn to help me forgive others who hurt me and, moreover, to help me forgive myself. I learned that forgiveness is not simply for the benefit of those who have hurt us, it is for ourselves, too. During Greenbreath, I also asked Hawthorn to help me open my heart. I could feel her unbinding, unshackling, and cracking open my heart space. Afterward, I felt a tremendous lightening in my being, as I had been given so much more space to breathe and to feel and to be. *I felt much more expansive. Another participant even commented on the shift in my being after Greenbreath, and I could even see it in my face—once deep-furrowed brows had lent themselves to a smoother forehead and a curious gaze. I am grateful and in awe of my own light. Hawthorn made me realize how much I missed* me. *Over the past month I've noticed many other shifts, too. I've been surrounded by more positivity and feel lighter. Just last week I noticed how even my clothes choices have shifted, as I want to wear only what makes me feel beautiful, soft, and flowing. I realized that I had been hiding behind the same dark, stiff, rigid garments for five years, subconsciously trying not to be seen and to not let others in. Hawthorn is teaching me that I do not need to wall off my heart to protect it, and she has offered her protection. It's like a horizontal airy protective hug surrounding me. I feel more like myself than I've felt in years. I feel a smile in my eyes and more ease and flow in life. Thank you, Hawthorn, for helping me unbind my heart and restore it to its rightful place.*

Long Journey from the Head to the Heart

Hawthorn is well known for its heart-healing gifts on all levels—physically, emotionally, mentally, and spiritually. In my book *Plant Spirit Healing,* I write about these different levels of healing:

> Hawthorn's greatest and most renowned gift is its healing relationship to the heart. On a physical level Hawthorn improves circulation, lowers blood pressure, and reduces palpitations and arrhythmia as well as serving as a remedy for degenerative heart disease and heart failure. As a heart tonic Hawthorn is preventative for all manner of heart conditions. As a flower essence Hawthorn relieves a heart saddened by separation or love loss. It also opens the heart, helping one to give and receive love. The spirit of Hawthorn can bring balance to the heart organ, the official of the fire element within the Five Element modality, and can also clear the heart chakra. But the most important benefit of Hawthorn in Plant Spirit Healing is the ability to put the heart back in its rightful place as the pilot allowing the mind to serve as the copilot.

This last piece, about the heart as the pilot, is so important because we have been indoctrinated into believing that our brain reigns supreme while our heart is relegated to the realm of sentimentality. The truth is that our heart is the primary organ of perception that makes the decisions and then sends them to the brain to carry out the decisions. An old Native American saying reveals the struggle to lead with one's heart: "The longest journey you will ever take is from your head to your heart." It is easy to get stuck in one's head trying to "figure it out." If we could let our heart do what it is designed to do—to be a good manager—we would be much more efficient and effective with our decision-making. This vital gift that Hawthorn offers is imperative to our ability to live in coherence, co-creative partnership, and unity consciousness.

In my Earth walk, I strive to be in service to the Holy Heart—that shared space of love that is greater than oneself. Sometimes this requires that I heal my heart from loss, which seems to be at the crux of all brokenheartedness.

As I travel through the alcoves of my heart, I dust the cobwebs off my musings where the echoes of heartbreak collide. I glimpse the worn track of former wounding and, like a feather, trace each curve and cavern. Will I ever heal?

I ask Hawthorn to fill my heart with her essence as I slip into her dreamtime. "Oh, sweet redeemer, hear my cry. See me, feel me, touch me, heal me."

The queen in all her majesty appears and wraps me in her cloak of golden radiance. The deep hole of loss where despair resides receives a minute drop of honey-like liquid of love and the healing begins.

Leaving a Legacy

During a Hawthorn initiation when we asked for Hawthorn to bring us a significant dream, I dreamt about my ovary eggs being harvested, and since I'm way past that age in my life, I knew it was a metaphor. After discussing my dream with the initiates, the meaning emerged. This was about what I was leaving behind—my legacy. The interesting aspect of the dream was that it was about the eggs that make children. The next piece of this understanding was that I am (we all are) leaving a legacy for the children. Are we visioning the effect that our thoughts and actions will have on the seventh generation? What kind of a planet are we leaving for even the next generation? These are big questions for us to ponder, and it is imperative that we take action now to ensure a livable planet for our children and their children. If we subscribe to co-creative partnership with Nature, which includes plants, then we can focus on the solution by first consulting with the plants (in this case Hawthorn) who give us insight into both the macrocosm and microcosm, so that we may proceed with a fuller picture of how to leave a thriving planet as our legacy (see color plate 11).

Lifting the Veil

I have found that if I don't approach Hawthorn with respect, then the veil descends and I'm not able to hear (or see) the gifts of Hawthorn. With time, I discovered that it is the queen of the fae who discerns what favor one will receive from Hawthorn. My friend and colleague Emma Fitchett (Farrell), in her book *Journeys with Plant Spirits*, writes: "The race of beings that Hawthorn concerns herself with are the fae. She is known as one of the most magical trees in the British Isles and a portal to the fairy realm. She is the guardian of the gateway into this dimension through which we can only enter with her permission." Emma suggests that the spirit of Hawthorn allows the fairy queen to live within her so that when we honor and respect Hawthorn, we are also honoring and respecting the queen of the fae because they are one and the same.

My former student Lauren shares her experience during the Hawthorn initiation.

Releasing the Heart

I arrived at Sweetwater Sanctuary for the Hawthorn initiation in early summer of 2023 with much anticipation and gratitude to be able to return to be with Pam and the plants in ceremony on her land. I have spent many days and nights at Sweetwater Sanctuary over the past decade working with the plants as conscious, dynamic living beings, and it is difficult to express the depths of transformation I have been gifted with through this approach. I have been led down a path toward integrity, healing, opening, liberation, connection, purpose, releasing the past and dreaming the future, and so much more. Any modicum of nonharm, and perhaps even usefulness, I can offer as a human being on the planet was born, shaped, and fired in the spiritual kiln that is Sweetwater Sanctuary, with Pam as a teacher and guide and, most especially, with the plants themselves.

I showed up to the Hawthorn initiation carrying a bundle of confused feelings that felt to be in the way of my ability to center in my own heart and my own energy. This knotted-up energy had been building and ripening ever since Hawthorn had come to me in a dream the previous fall. I felt whipped around by my own emotions, other people's stories, the intensity and exhaustion of caring for my two small children, the staggering layers of crisis and trauma unfolding on Earth, and uncertainty about how to be in service.

We opened the ceremony and recorded our intentions on a piece of paper, which we slipped underneath the altar cloth. Here is part of what I scribbled down: "My intention is to ask Hawthorn for clarity and come deeply into who I am and the work I am here to do. My own soul vibration and purpose and—how to support myself and grow my power to be in service to healing for Earth and all beings."

We then prepared our Hawthorn elixir, which we would drink together eight times through the weekend. After taking our first drink, we packed up blankets, notebooks, and chairs for a walk down the road on pilgrimage to spend a couple hours sitting with a Hawthorn tree in a nearby field.

Often, when I approach a plant or tree to sit with them, it takes a few moments to quiet myself down and tune in. The communication I engage with from plants and trees can be subtle and unfold over time as feelings, sensations, colors, sounds, and memories. This is not what happened with Hawthorn. I walked up to the tree, a bit sweaty from our walk, focused on avoiding thorny branches and juggling my belongings. As I meandered around to find a shady spot for my blanket, I heard a strong voice command, "Write!" It was such a loud and clear directive that I looked up, a bit startled. I realized that the voice belonged to Hawthorn, reaching out to me before I could even sit down.

I quickly spread out my blanket, opened my journal, and, somewhat shyly, allowed my hand to write what came:

Hi. You need to be shown the way to the heart. It is time to know how to find your way home. I can be your guide so you do not have to be in charge and you can get more rest. All the trees can calibrate you. You have been lost and asking about resonance. You have not been knowing the right vibration. So many possible vibrations to calibrate to. Trees will always bring you back to the home vibration. You need to hold space open—wide open—for your soul to come through. Your soul is extremely powerful, and you have forgotten aspects of this.

I asked: "What does this look like?" The reply: "Constantly checking in with your heart until this becomes a habit."

When we returned from our pilgrimage, we gathered in a circle for a drum journey to visit with the spirit of Hawthorn. After traveling down into the underworld, I met with the spirit of Hawthorn and said spontaneously, "I need to be better to other people." Hawthorn replied, "Yes, that is true," and then said, "But we need to get rid of these thoughts!" I felt a tingling sensation in the left side of my head, and then Hawthorn put me into a sleep state. I was still listening to the drum, but I was sedated without much awareness of what was going on until it was time to come back from the journey. As the drumming stopped and I sat up, blinking my eyes open, I felt shifted. I could sense that Hawthorn had pulled to the surface something from deep within my inner realms.

I opened my journal and wrote: "So much confusion and self-doubt swirling, deep feelings of inadequacy and confusion—my own participation in this harmful way of living on the planet. So many choices I am making to perpetuate the old."

As the afternoon waned we made our way up the hill to walk Sweetwater's beautiful seven-circuit stone labyrinth. An adolescent Hawthorn tree grows at the southern edge of the labyrinth, their branches sweeping out over the edge of the outer ring of stones. Three of us walked first into the labyrinth, stationing ourselves at various points so that we could ask a contemplative question to each person that came in to walk after us. I walked halfway through and stood at the edge of the path so that I could ask the question that Pam had assigned to me to each person that followed: "How will your heart sit on the throne?"

I spoke these words out loud over and over, meeting the eyes of each person. After the last person had walked by, I turned to finish my own walk. As I neared the center of the labyrinth I clearly heard the answer:

By releasing judgment.

A shiver ran through my body and my eyes fluttered. "Oh. Yes. Thank you, Hawthorn."

I felt electric, sensing that I had just been handed a critical piece of information. I knew then that during the drum journey, where I had been sedated, Hawthorn had extracted and pulled to the surface thoughtforms of judgment. These thoughtforms were now floating at the surface of my energetic field and were coming into view. The knotted tangle of energy that I had brought to the initiation and asked Hawthorn for help with was starting to loosen. Hawthorn was digging in. I could see and feel that these thoughtforms were something that I was tightly holding but that were not native to me. As we gathered back in circle for our final drink of the day, I said out loud, "Hawthorn is retensioning my energetic system. Showing me what is me and what is the junk that needs to go."

I went to bed Friday night following instructions to ask Hawthorn for a dream. I did receive a dream: a series of different scenarios played out where my children were in danger, and each time when I felt the paralyzing fear that I was going to lose them, they made it through the ordeal and were fine. The dream was a tender gift to help me to release fear. Early in the morning of the second day, I woke up and immediately wrote down these words:

Judgment. Fear. Wanting and needing to claim who I am, needing to express and offer my gifts but also needing to not operate from a place of ego.

Yes—Self-esteem, self-confidence, humility

No—Judgment of myself and others

I ask: "Hawthorn, what does this look like?" Answer: "Letting go of fear. Identifying judgment thoughtforms."

Through the morning of the second day I was hyperaware of the stream of thoughts running through my mind. The wires of my brain felt crossed. This whole bundle of thoughtforms that Hawthorn had brought forward were shouting and muttering, half unplugged, dramatically—almost comically—flailing around. I was constantly having judgment thoughts about myself, another person, or the world, one after another. Each time one would come, Hawthorn would say with gusto, "There's one! There's one! There's another one!"

At midday I was standing quietly by the bubbling stream outside the center building, when I heard Hawthorn say, "OK, now I want you to feel into the physical space of your body. Find out where these thoughtforms live and how they are being held in your muscles and bones."

I scanned my body. I found that these thoughtforms were knitted together into a hard scaffolding around my heart. I could see that for many years I had been constricting the movement of my arms, chest, and back, holding my elbows tightly against my rib cage and collapsing my shoulders forward over my heart. As I breathed into this newly discovered tension, I could see just how much of my life force was tied up by holding this cage in place. I was doing this to keep my heart safe from feeling too much, from being annihilated by painful experiences of life. My judgment thoughts were guarding the door to my heart, keeping out emotions I thought I could not handle.

At midafternoon on Saturday, we took another drink of Hawthorn elixir. Then we took a break, and I went to my tent, lay down, and fell into a deep, heavy sleep. Hawthorn's herbal actions as a nervine relax the body and, I was learning, the whole energetic field. When the field is tightly wound together, it is difficult to let go of what is not needed and rework the system.

I woke up with a jolt after half an hour, drool all over my pillow, and ran back to the circle. We took another drink of elixir. After sitting for a while with this next drink of Hawthorn working in my physical body, still half in a dream state, I said out loud to the group, "Now I am feeling the grief that is underneath the judgment." In that moment the grief was wispy, barely identifiable, but it was starting to bubble up through the cracks that were forming as the cage around my heart was beginning to come unbound.

Our afternoon activity was to make Hawthorn masks to wear for a Council of Hawthorn to be held around the fire later that night. Rain pattering against the windows, I sat in my seat sewing a soft green velvet fabric, cutting the top edge of my mask in the shape of the irregular jagged edges of the Hawthorn's leaves. Hawthorn had told me on the first day that their leaf edges were shaped like an EKG wave and mirrored our heart rate's healthy variability. I felt so drawn to the softness of the fabric, the thorn-like needle as a tool, the simplicity of the green, the quiet timeless act of sewing. Through this rhythmic stitching, I felt lulled by Hawthorn, deliciously calm, far away from my thoughts.

After finishing our mask making, we returned to the circle, wearing ceremonial clothing, to take another drink. Dressed in my blue dress and white shawl, my hair braided and pinned around my head, I sat for a moment with the elixir in my hand before I took a sip. I brought the drink to my lips. As Hawthorn entered my body, I felt a strong, unmistakable vibration in my sternum, as if my chest had been plugged into an electric current. Hawthorn, whose voice by this point I was getting used to hearing as clearly as my own thoughts, exclaimed exuberantly, "Ah, yes, now is the time! Stay present for this!"

My rib cage began heaving and shaking. My first instinct was to clamp down and cut off the sensation because I was sitting in a circle with eighteen other people and I did not want to make a scene. But I also knew that something big was happening. The work of the initiation had been building, enough healing energy had been moving, and, like floodwaters behind a dam, like a mother and unborn baby whose time has come for birthing, it was time for a bursting, and Hawthorn was now getting into the driver's seat of my body. The train had left the station and was roaring down the track.

The vibrations escalated, and I slid off the low pillow I was sitting on and thudded onto the floor, my body overtaken. I began to let out sounds, most easily described as sobs, that were emanating from the center of my heart. As they welled up I again felt the urge to shut the process down— "Can I really make these sounds in front of all these people?"

All this shaking, rattling, heaving, and sobbing were crumbling the cage around my heart. Chunks were falling off and crashing to the ground. My red beating heart was pounding and bursting and shuddering awake and oozing outward. I felt incredible sorrow. I was suspended in the heart of grief. Hawthorn was practically shouting at me, "Here it is! These are the feelings that you have been afraid of and walling yourself off against. This is what it feels like! This is grief! This is life! Now you know! Now you can let go of this cage of fear and judgment that has been your heart's prison!"

Like a sea creature who is flung out of the ocean onto the sand, flopping and gasping, fearing it is about to die but who then realizes it can breathe air and is meant to live on land, my heart—cradled by the tenderness of Hawthorn and the circle of souls around me, Pam holding my head in her lap and stroking my hair—was taking its first breathes of true grief and true love. Wobbling like a colt on the shores of truth. As my body calmed down and my breathing settled, I said out loud, "It's real." In that moment, what I knew to be more real than anything I have ever known is that the deepest grief is the deepest love. They are one, and this is what the heart is built for, and this is what is real. This is the domain, power, and skill of the heart. Finding my way home to live in these heartlands is why I am alive on Earth at this time.

Hawthorn's Healing

Hawthorn is a powerful healer especially for the heart, which was mentioned earlier. (***Note:*** If engaging in a Hawthorn initiation, please be aware that Hawthorn may interfere with beta blockers. If this is a concern, an essence of Hawthorn can be taken internally instead.) What I have noticed in my years of working with Hawthorn is that her healing goes deep and is progressive. Because of the heartbreaking milieu we are currently living in, our hearts constantly need healing. Hawthorn's heart-healing gift *continually* provides healing. Dianne who attended the Hawthorn initiation says, "the three-day immersion was the tip of the iceberg," meaning that the massive healing that took place during those three days was only the beginning. Hawthorn carries the gift that keeps giving (see color plate 12).

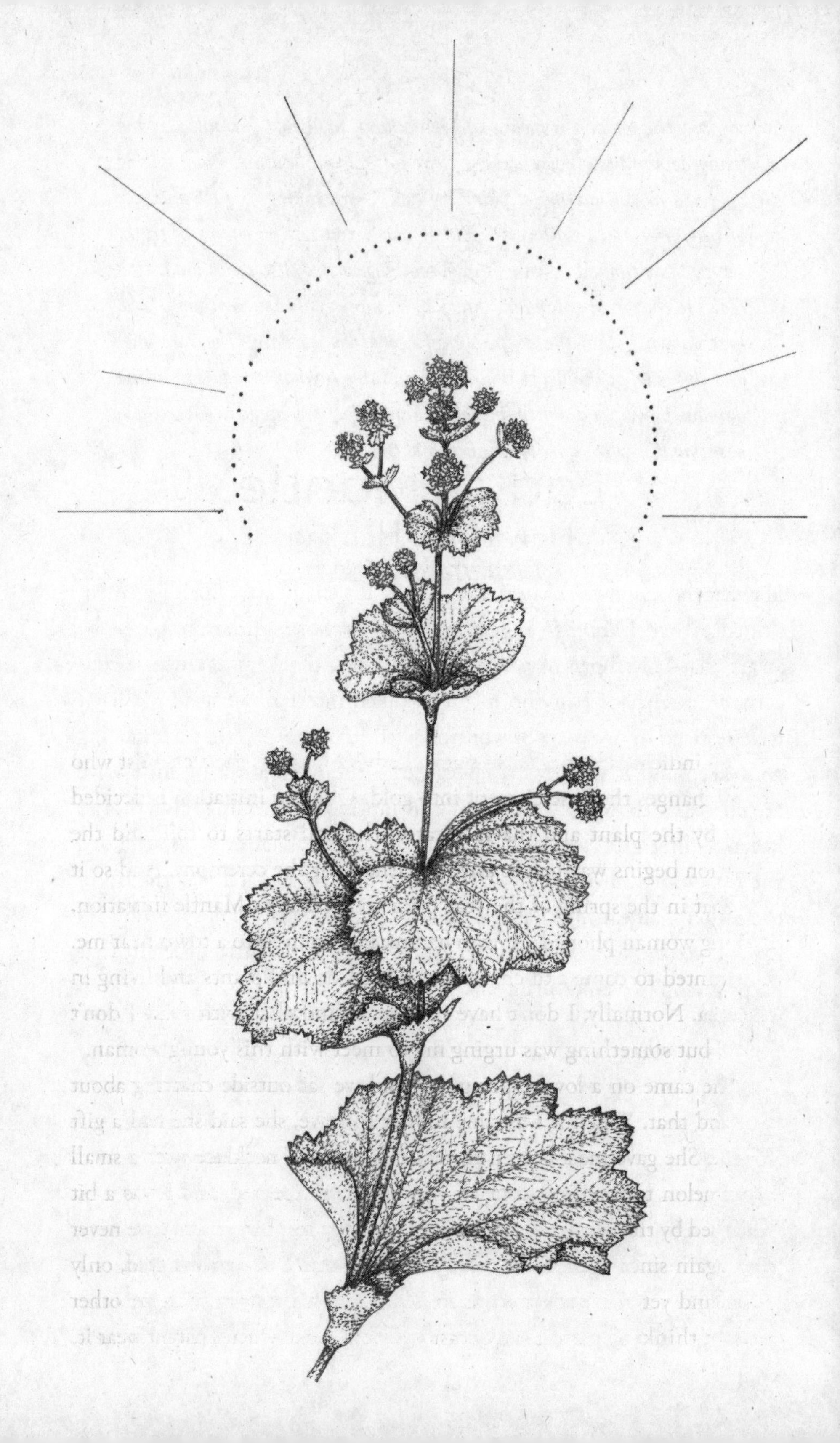

THIRTEEN
Lady's Mantle

Alchemilla vulgaris

As indicated by her Latin name, Lady's Mantle is the alchemist who changes the base element into gold. Once an initiation is decided upon by the plant and the facilitator, the ball starts to roll, and the initiation begins way before anyone gathers for the ceremony. And so it was that in the spring of the year prior to the Lady's Mantle initiation, a young woman phoned me who had recently moved to a town near me. She wanted to come and chat about what I do and plants and living in the area. Normally, I don't have time to sit and chat with folks I don't know, but something was urging me to meet with this young woman.

She came on a lovely spring day, and we sat outside chatting about this and that. When it was time for her to leave, she said she had a gift for me. She gave me a beautiful, very delicate gold necklace with a small watermelon tourmaline stone. It was quite unexpected, and I was a bit surprised by the gesture of someone I had never met before and have never seen again since then. Even more curious is that I never wear gold, only silver, and yet this necklace was so alluring. I put it away with my other jewelry, thinking perhaps an occasion would come when I might wear it.

The next morning I was making my usual walkabout through my gardens, greeting all my plant friends, when Lady's Mantle stopped me and, clear as a bell, the touch I received translated as, "Where is your gold necklace?" (or was it, "Wear your gold necklace"?). I was stunned and then realized this meeting with the young woman was part of the Lady's Mantle initiation, and she brought me the gold necklace to wear as a way to bring the gold frequency into my life. I immediately fetched the necklace and wore it throughout the time before, during, and after the Lady's Mantle initiation.

The Gold Frequency

Lady's Mantle is bringing forward the gold frequency, which is that of the New Earth, when a golden age of unity consciousness becomes manifest. This is a time when love and compassion flourish, peace and justice prevail, and Nature is experienced as kin. The golden light of Lady's Mantle reaches into the recesses of our being and lights the way to this era. Marie shares her experience of the gift of Lady's Mantle.

> *Green hands held and surrounded me with support and brought light through my insides, primarily through my womb space, which was darkened and heavy. Often when I'm creating, I follow the flow and was surprised to find the bottom layers of this image looking more and more like gold nuggets emerging from where I held fear, tension, and pain in darkness. I'm finding* Alchemilla *guiding me deeper and deeper into a heart-centered way of being. I'm learning this is not the absence of darkness but more an emerging, with gentleness, of golden light from those darkest spaces within myself (see color plate 13).*

Feminine Rising

Lady's Mantle is a well-known herb for supporting women's health, especially reproductive issues, so it is not surprising that on a spiritual level she would support the feminine rising, which is a growing awareness of the need to revive the feminine principle, being acted upon by women and men alike. (***Note:*** If engaging in a Lady's Mantle initiation, please be aware that it is not recommended for pregnant or breastfeeding women to ingest Lady's Mantle. Instead, an essence of Lady's Mantle can be taken internally.) Cultural mythologist and theosophist Carol Wolf Winters writes in her book *Who Said, "God Said"? The Truth behind the Myth of Female Inferiority*:

> It all started long ago, at the dawn of human consciousness, when some said, "In the beginning was Mother Earth, the primal vessel that contains all things." The Great Mother was inclusive. From her womb emanated all life, and from her body all of her family received the gifts of nourishment, shelter, and transformation. When her children died, she enfolded them back into herself to be reborn anew. This early concept of what was later to become the feminine principle was that of a nature-based, interconnected existence for all creation, both in life and in death. Eventually our ancient "fore parents" understood that the female body also embraced the creative life-giving patterns of Mother Earth: its womb generated and protected life, its breasts nurtured, its arms embraced and comforted. The feminine principle became associated with early female experience, and was conceived as the creative vessel of life that contains, nurtures, and protects.

The pre-feminine-principle-concept era that Wolf Winters describes as "nature-based, interconnected existence for all of creation" has been relegated to the far reaches of the human psyche. The notion that all of creation, including humans, could actually live with interbeing has fallen into the abyss of amnesia as the current consensus reality promotes the paradigm of us and them, where separation from self, other, and Nature reigns.

At the core of the initiation process are the plants and trees that are stepping forward to help us remember our true nature and how to live within its essence. In this case, Lady's Mantle is bringing forth her gifts of helping us to recognize the feminine principle and its life-giving qualities while encouraging us to engage in the feminine rising, like the phoenix ascending from the ashes.

The following is a letter written by Lady's Mantle to April's future feminine-rising self.

Dear Future Self of the Rising Feminine,

You are like a spring lifting up the secret waters of the deep to finally shine in the sunlight.

You are whole and unashamed. You joyfully greet the morning, the evening, and the stars.

You tend the fires of life with so much loving care. You make soup, delicious and abundant, enough for everyone.

You have a wide and comfy lap as well as a springy step. Arms that embrace and lift and sway and rise up toward heaven.

Your seeds flourish, your words nourish. The stories you tell draw healing into the world.

You find the small gift that shifts everything toward hope.

You dance your offerings, you join your sisters in song.

You gather the children and protect them.

You stand strong. You weep without restraint.

Your life becomes a garment of mercy, woven thread by thread of love. These threads become more finely spun, more tensile strong, as years go by, until they are both nearly invisible and breathtakingly beautiful.

Blessings to you dear one,
The Lady

April's simple reply to Lady's Mantle is: "You hold a light up to my present, searching self, and with a smile you assure me that all is well."

Transformation

Metaphorically speaking, alchemy is the journey of evolving into our greatest potential or awakening to your true self. I experience alchemy as a magical expansion brought forth by transformation.

After the Lady's Mantle initiation was over, her gifts continued to unfold, and I found myself in Belize for the winter. Shortly after I arrived, the full moon in January took place (see color plate 14), which, with Lady's Mantle's impetus, triggered a massive shift in my being. I found myself mining the depths of my being physically, emotionally, mentally, and spiritually—all levels. I explored the subterranean landscape of my soul, where old, nonserving habits, patterns, beliefs, and attitudes were stuck as if in quicksand. One by one, I dug them out as the stain of grief demanded I recalibrate. Each was unpacked, opened, and dissected to understand their inner workings. Then I left them to ferment until all that was left was a rich compost of decaying grief and loss where I planted seeds of beauty, making possible something new to flower. This alchemical transformation was truly a miracle.

After Rhiannon had arrived home from the Lady's Mantle initiation, she described how she felt: "Since the initiation, I feel I have been clearing out a lot of old patterns that are coming up in ways they haven't in a long time. It's not always comfortable but Lady's Mantle has helped me to gentle into it."

The Protective Mantle

The Lady in Lady's Mantle is the Virgin Mary, and the mantle is her cloak, which serves as a protective wrap. Lady's Mantle offers protection when there is an issue of violation of boundaries, especially for women and their bodies. Within our current social landscape, protection is needed constantly. During the Lady's Mantle initiation, a vision of deep transformation emerged from within the group, where protection from "other" was no longer necessary.

Myra describes this vision: "We recognize we are one of the distinct expressions of life within the circle, and there is no need to defend, no need to protect, and that in itself is the great repair of the tear that has created separation in how we and other life-forms move with that which sources life."

If Lady's Mantle's gift of protection can extend to envelop the vast web of life so that *all* life is under her mantle, it is no longer necessary to hold onto the need to "protect against." Protection becomes omnipresent, and our state of being is girded by safety and sanctuary, engendering trust and ease.

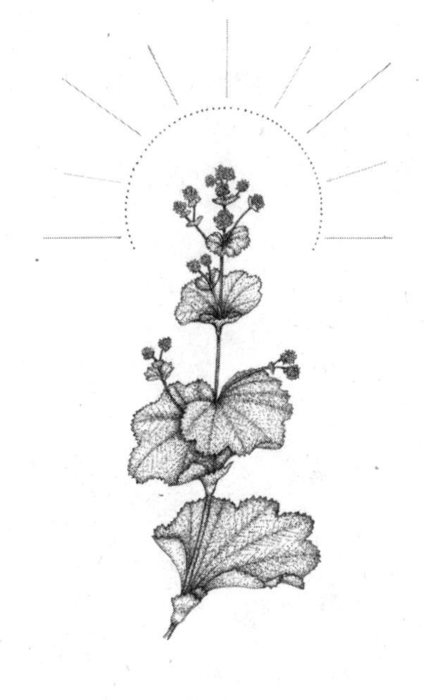

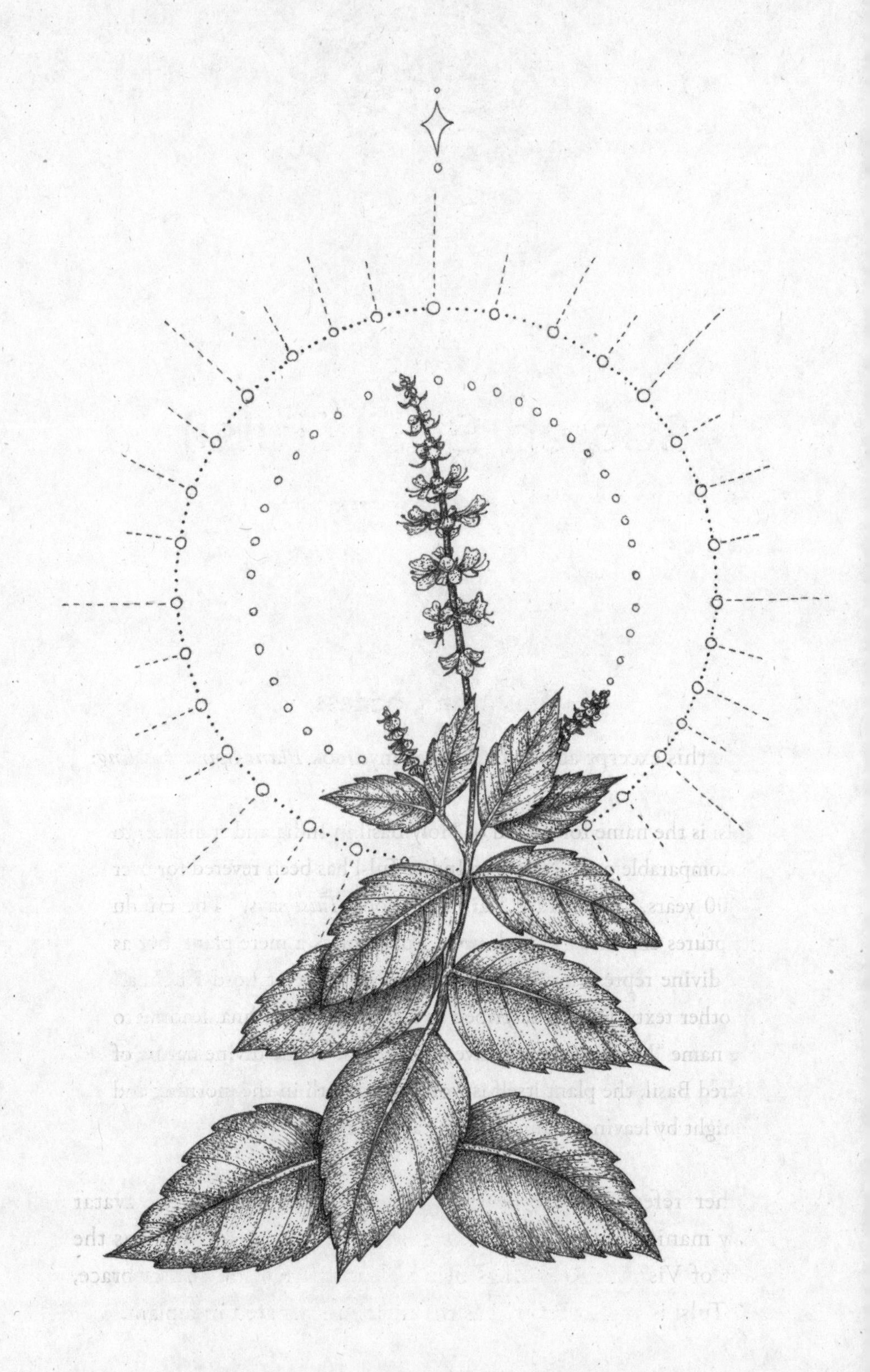

FOURTEEN
Sacred Basil (Tulsi)

Ocimum sanctum

Tulsi, the Goddess

I share this excerpt about Tulsi from my book *Plant Spirit Healing*:

> Tulsi is the name for Sacred or Holy Basil in India and translates to "incomparable one." Native to India, Tulsi has been revered for over 5,000 years, and as Yash Rai in his book *Tulsi* says, "The Hindu scriptures enjoin us to look upon Tulsi not as a mere plant, but as the divine representative of the God Vishnu or of Lord Krishna." In other texts Tulsi is described as a consort of Krishna, leading to the name "Mother of the Universe." Because of the divine nature of Sacred Basil, the plant itself is worshipped both in the morning and at night by leaving a lamp burning at its base.

Other references are of the goddess Tulasi who is the avatar (earthly manifestation of a god or goddess) of Lakshmi, who was the consort of Vishnu. Regardless of which interpretation you embrace, clearly Tulsi is highly revered as the divine incarnated in a plant.

Liddy shares her experience with Tulsi as a goddess.

When Tulsi showed up as the goddess that she is, I invited her into my medicine wheel. She sat opposite me and brought her third eye to my third eye. What happened then can only be described as a moment of enlightenment. I felt the vastness of the universe and my connection to it. I understood that this is who I am, who we all are. She has been with me ever since.

Caroline describes her experience of reverence in the presence of such a divine being.

Tulsi is helping me explore sacred devotion—what it is and how to be or do it. In a shamanic journey with Tulsi, my Wise One made me wash my hands, sing a song, and instead of going to Tulsi, we just sat in prayer and meditation. Eventually, Tulsi arrived—first, as a piercing sensation in my heart with warmth and then as a goddess in a flickering sari-flame. When asked what gifts I could bring to her, Tulsi said, "Your attendance, presence, devotion." This invitation has been revealing to me my own wounded relationship with devotion and attendance.

LeAnne experiences the magnanimous spirit of Tulsi and describes it in this passage.

Tulsi is the resurrection of spirit/my spirit, ready to aid and unleash the tightened specks of the nothingness within. She has helped me to understand that she is the elixir of all life, filled with pure and sacred truth. Now that I have experienced her medicine I will never open up the same door nor see myself as the same person again. She is the elixir of infinite knowing. She unlocks the parts of my soul that have been hidden since I can remember. She diffuses the dark parts of me (perhaps the traits of my ancestors) and calms the beast within. Experiencing her medicine gives me permission to trust myself and to unlock my inner spirit of my sacred and true nature.

For me, while sitting with Sacred Basil, her vibratory resonance is felt not just in my head but more like a halo in my auric field. Along with this glistening halo is movement that is like swooning along with her spicy aroma. She's so much more than a goddess; she's All That Is. The felt sensation of "swooning halo" invokes the emotion of exquisite joy that bubbles up until it explodes into ecstasy. The ecstatic journey takes us out of static time and places us in the Now moment of no beginning and no end, only the present that is eternal. Touching the ecstatic (is this why people do drugs?) to experience the complete freedom of timelessness connected to All That Is. And yet, it's all around us at any given moment, this ecstatic moment. Thank you, Tulsi, for sharing your gift of touching the Divine.

Soul Liberator

From your lofty locus, Tulsi, you have access to that which never dies—one's soul. You carry the gifts of retrieving soul parts that go missing, and repairing and renegotiating soul contracts between primary soul companions (usually family members). You have access to the blueprint of one's soul and the ability to interpret the soul's design. Here, in this space where all is connected, you illuminate the many aspects of the soul, including what Martín Prechtel refers to as the "indigenous soul." During a Tulsi initiation, I wrote the following letter to my indigenous soul:

Dear Indigenous Soul,

I miss you, I need you. I long to walk the mountains and valleys, along the streams and through the forests with you. I want to sip from your cool waters with moss entwined in my hair and red clay adorning my cheeks while resting in the comfort of your deep embrace. Let me leave the computer, cell phone, and all those devices behind because what will become of me if I turn away from you and serve only the machine gods? Will I fall so deeply into amnesia that I will not be able to hear your whisper calling me home? I want to drink from you, feed from you, play with you, grieve with you, love with you, touch the ecstatic with you, and remember who I truly am, remembering the time when I knew none other than you.

In an interview with the *Sun* magazine, Martín Prechtel further describes the indigenous soul:

> We are all still human beings. Some of us have buried our humanity deep inside, or medicated or anesthetized it, but every person alive today, modern or tribal, primal or domesticated, has a soul that is original, natural, and, above all, indigenous in one way or another. The indigenous soul of the modern person, though, either has been banished to some far reaches of the dream world or is under direct attack by the modern mind.

In *Secrets of the Talking Jaguar* Prechtel also writes: "For there to be a world at all, every indigenous, original, natural thing must start singing its song, dancing its dance, moving and breathing, each according to its own nature, saying its name, manifesting simultaneously its secret spiritual signature."

Thank you, Sacred Basil, for helping me to know and remember my indigenous soul.

One of my students, Suneet, describes how Tulsi has touched her especially on a soul level.

> *My journey with Tulsi has been intricately woven into my life and experiences over the last few years, to the point where it has been difficult to sit down and tie it all together without embarking on a long and winding tale.*
>
> *I first met Tulsi through an aunt who showed me tea bags and told me how good Tulsi was for the body. I was young, more interested in enjoying myself, and took little notice. But the memory stayed with me, and at the time I could feel a deeper chord being struck. It wasn't until years later that I reconnected with Tulsi through Pam Montgomery's apprenticeship. We had just completed Part 1, and at the end of Part 1 week, Pam said that the following year, in Part 2, we would be working with Tulsi.*
>
> *However, two months later my mum came home with a Tulsi plant that had been gifted to her at the temple. Being the only Indian in a group of thirty-plus students, I knew instantly that this was hugely significant for*

me personally. Also that my mum—in her life—had never ever given me a plant. It was here I truly began my journey with Tulsi. I nurtured and loved this plant: it was so small. But a gray winter in the UK meant that it didn't last very long.

The week before the Tulsi plant died, I was running a Mugwort ceremony and someone gifted me another Tulsi plant and this plant is still with me today. I was even more careful to take care of this plant, to nurture it through the winter, hear its needs, and generally give it love. I found myself being very tactile with the plant, giving it little kisses. This process taught me so much about devotional love. There was no reason for me to love this plant so dearly; I hadn't had any magical moments with Tulsi yet. But Tulsi had already found her way into my heart. I cared for Tulsi with no expectation of reward or reciprocation.

As the months went by I found myself in a healing session with a client, and as I was working energetically to clear her body, an "unfriendly" energy came out of her—and then latched straight on to me. I tried everything to shift it, and in the end I just silently sent out a call for help to the spirit world. In a flash of bright pure light, Tulsi arrived and booted the energy out, and I recognized in that moment that Tulsi reminded me I am a Divine being, a person who carries a soul that belongs to God, and nothing else can latch onto that. It is a lesson that I still carry within me now.

Tulsi began working with me in other ways, healing two very badly damaged soul parts from past lives, slowly, carefully, each new step bringing me closer to feeling more whole. Tulsi taught me about my soul's purpose and soul essence. And if that wasn't enough, the biggest, most profound healing I have had with Tulsi is healing my relationship with my bloodline. Tulsi helped me heal displacement from my motherland, my own personal challenges with things in my culture and religion I did not agree with, and the shame that is placed on a person being different and brought me back to my ancestral medicine of Ayurveda.

To me, Tulsi is a master plant, a life-walk plant. Its spirit brings me continued teaching, growth, and learning, but most importantly for me, it is a plant I love to love.

To Adapt Is to Thrive

Plants and trees have been adapting to a changing environment since they emerged from the ocean some 400 to 450 million years ago, and because they have continued to evolve for millennia, they are mavens of adaptation. They have so much to teach us about how to adapt during times of massive change, like the one we are in the midst of during this modern era. It turns out that there are certain plants that possess supreme qualities of supporting adaptation and, in the world of herbal medicine, are referred to as adaptogens, which Tulsi is considered to be. Even though these adaptogens are spoken about mostly in regard to their physical healing properties, the attributes that give them this designation can be extrapolated to serve as guidelines for adaptation on a larger scale.

The following explanation of adaptogens is a paraphrased excerpt from *The Rhodiola Revolution* by Richard Brown and Patricia Gerbarg. In ancient healing practices, there was a group of herbs considered to be "elite" or "kingly" because they were known to increase physical and mental endurance, reduce fatigue, improve resistance to disease and enhance longevity. In 1948, Dr. Nikolai Lazarev and Dr. Israel Brekhman, with other researchers from the Siberian Academy of Sciences, tested 158 herbs known in folklore to be super herbs. They found that certain herbs support the healthy function of every system in the body and offer protection from biological, chemical, environmental, and psychological stressors. Lazarev coined the term *adaptogen* to describe this class of herbs. He and his team of researchers established three fundamental criteria to identify an herb as an adaptogen. The following list is from the book *The Rhodiola Revolution*.

- **Nonspecific resistance.** The herb must increase the body's resistance to a broad range of agents, including physical, like heat, cold, and exertion; chemical, like toxins and heavy metals; and biological, like bacteria and viruses.

- **Normalizing action.** The herb should normalize whatever pathological changes or reactions have occurred, meaning it brings balance to the condition, steering it toward normal function.
- **Innocuous effects.** The herb must cause a minimal, if any, physiological disturbance or side effect and be very low in toxicity.

Since Lazarev's time, much work has been done to illuminate the properties of adaptogens. The current understanding and working definition of a true adaptogen is an herb that works on the endocrine level to modulate the body's response to physical or emotional stress, increasing a person's ability to maintain optimal balance. Adaptogens can downregulate the hormonal cascade of adrenaline, norepinephrine, and cortisol that make up the stress response, preventing an overreaction that can damage and deplete the cells of vital energy, while at the same time they tonify and build the body's innate energy. Also, a true adaptogen has a balancing effect on the body's regulating systems, including the cardiovascular, immune, and neuroendocrine systems.

Because Tulsi possesses many of these adaptogenic qualities, on the physical level she gives us greater vitality and the ability to regenerate and lessens the depletion caused by stressors, mirroring her gifts on the soul level of restoring lost soul parts, calling in our original blueprint, and reminding us of our divinity. Because of adaptogens like Tulsi, we are able to not only survive through this tumultuous era of transformation but we can also thrive and embody true wellness in our body and wholeness in our being.

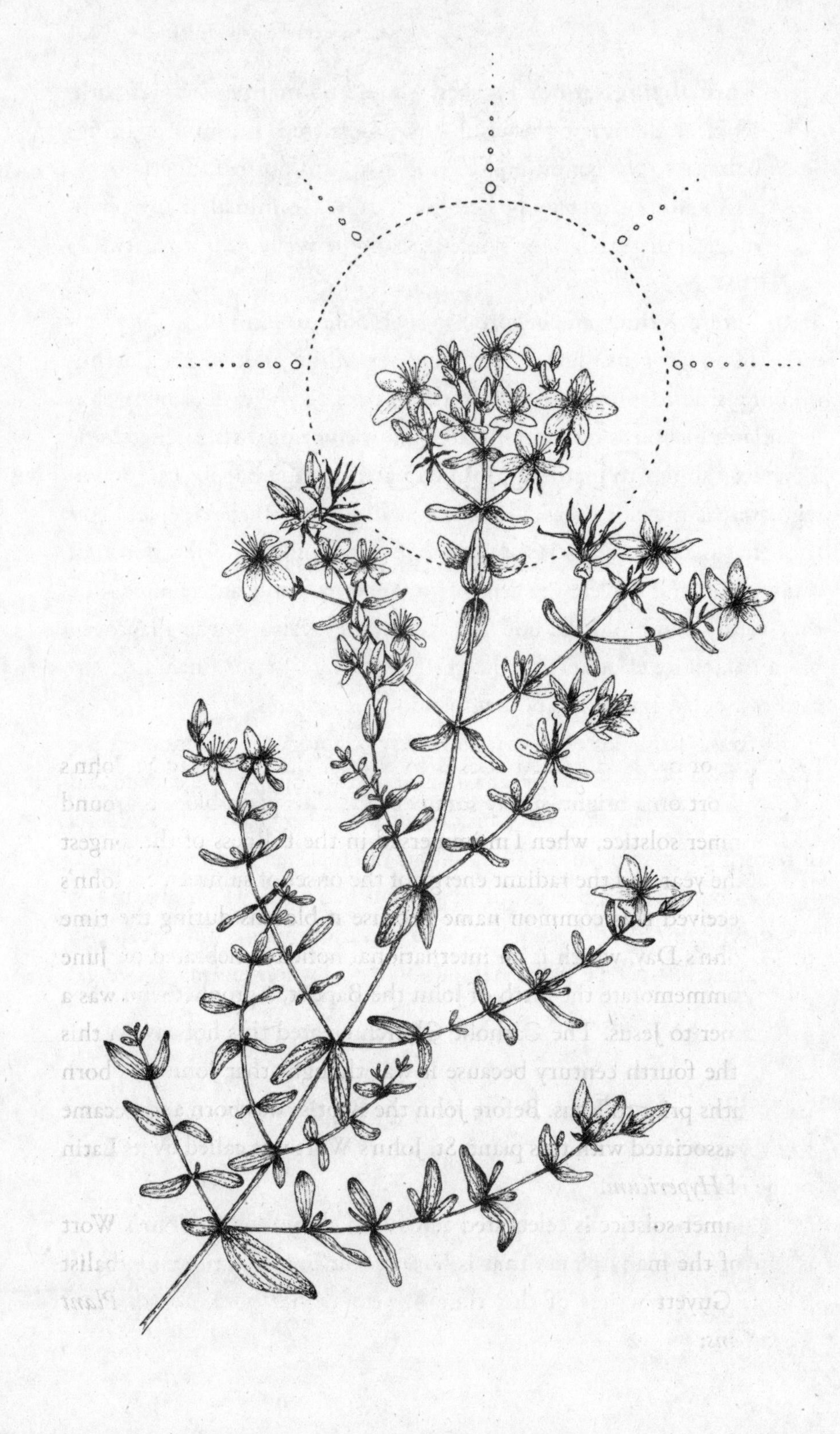

FIFTEEN
St. John's Wort

Hypericum perforatum

One of my favorite activities is to harvest the flowers of St. John's Wort on a bright, sunny summer day. This plant blooms around the summer solstice, when I'm immersed in the fullness of the longest day of the year and the radiant energy of the onset of summer. St. John's Wort received this common name because it blooms during the time of St. John's Day, which is an international holiday celebrated on June 24 to commemorate the birth of John the Baptist, a prophet who was a forerunner to Jesus. The Catholic Church created this holiday on this day in the fourth century because it was thought that John was born six months prior to Jesus. Before John the Baptist was born and became forever associated with this plant, St. John's Wort was called by its Latin name of *Hypericum*.

Summer solstice is celebrated across the globe, and St. John's Wort is one of the many plants that is honored during this time. Herbalist Carole Guyett writes of this time of year in her book *Sacred Plant Initiations*:

The Summer Solstice is a time to celebrate what has manifested. We celebrate the abundant growth that has occurred in Nature as well as the attainment and fulfillment that has occurred in our lives. It is time to enjoy what we have and who we are. We fully express our uniqueness and we call on the strength and power of the sun to charge our intentions. At the same time, the Wheel of the Year is turning and from now on the days will shorten. A transformation is taking place. We prepare to return to the darkness of the inner worlds, activated and strengthened by the peaking energy.

Historically, as the festivals of summer solstice covered the landscapes of the Northern Hemisphere, they took place on what was referred to as midsummer's eve. Carole describes the celebrations:

> In the past, people stayed up all night on Midsummer's Eve to watch the sunrise and celebrate the longest day. Bonfires were lit on hills at midnight and people danced around the fires and leaped the flames to rid themselves of bad luck and to assure abundance and the fertility of the land. Livestock were blessed with blazing herbs from the fire and flaming torches were taken sunwise around the field and home. Sacred trees were dressed with flowers and ribbons and it was a time of games and feasting. Sun Gods and Goddesses were honored, such as the goddess Áine in County Limerick, Ireland, where today people still pay homage at her sacred hill.

Matters of the Heart

St. John's Wort was used for divination during these festivals especially for matters of the heart, both relational and otherwise. My student Diana describes her love affair with St. John's Wort, which began during a St. John's Wort initiation.

> *This initiation was the most profound experience of my life next to the birth of my child. This initiation and St. John's Wort let me know that I was*

wanted and I was loved. I carried a core wound my whole life of not being wanted by anyone. I can feel it as I write this—the emotion that comes up when I think about how much space that feeling of being discarded and worthless took up in my body and how that distortion informed the choices that I had made in my life. The things that I settled for before St. John's Wort were not always pretty because when you feel completely unworthy you take whatever scraps you can get, physically, emotionally, mentally, and spiritually.

St. John's Wort was a gentle awakening to the green world. When I sat with the plant, the experience was one of soft waves of energy and beautiful colors; everything felt like it was going to be OK. The plant met me with love: I can still see myself lying across the bridge with the brook bubbling underneath, my head facing St. John's Wort growing along the bank, while I placed the flowers and leaves of the plant on my body and experienced its gentle support. I explored the energy of a new friend—looking at its form, learning the leaves and where it liked to grow—and being met along the way with love, trust, truth, and compassion. When we moved into the Greenbreath, St. John's Wort rebirthed me. I experienced being Earth, birthing everything, feeling the love a mother feels for her creations and truly knowing, in my bones, that everything on this planet is loved and wanted. St. John's Wort then had Earth rebirth me.

I walked away from that initiation knowing, feeling, and experiencing in my core what it is to be loved, to be wanted and understanding that is the truth for all of us and everything else is distortion. This plant initiation and St. John's Wort lives within me. I carry this medicine in the world while being blessed by and walking with St. John's Wort. I am the ambassador for this being in the world. When I'm with others, they receive the knowing of being wanted and loved because that is what I carry forward through this experience with St. John's Wort because I can never unknow this truth. There is knowing in the mind, feeling in the body, and then there is truth that is unwavering because the frequency of it cannot be denied. Plant initiation and St. John's Wort, specifically, brought me a truth that I needed and that cannot be denied. This green being taught me through the experience of the initiatory process that we are all loved, we are all wanted, we are not mistakes. It is through the grace of the living beings on this planet and the green beings, specifically, that exchange our carbon dioxide for oxygen, that feed, shelter, and clothe us, that we have the miracle of our very life. This miracle consists of love and that is St. John's Wort and what it brings into the world. I am eternally grateful.

Bearer of Light

St. John's Wort is well known to be a light-filled plant that dispels darkness when one becomes overly consumed by gloom. When I have experienced my spiritual fire burning low, I call upon St. John's Wort to rekindle the flame and to fill me with its radiant light. I imagine a pitcher of light being poured into the top of my head, creating rivers of light streaming through my being, touching every cell and reaching into my core. I shine brightly when I am filled with this light, and it emanates through me to all who I come into contact with, both human and nonhuman. When you are able to share light with the world because you carry enough to be luminous, Earth and all her beings benefit.

One of my students, Barbara, says that during a journey with St. John's Wort she was told that "it is my responsibility to shine as brightly as I can and to bring that to others." St. John's Wort not only provides light to replenish our depleted spirit fire, but it also instructs those of us who have chosen to be in co-creative partnership to bring this light to the wider world. Barbara continues, "I had an expansive consciousness inside my heart and head that was as bright as the sun. I had become a hollow bone filled with light and goodness bringing great hope for the future."

Carole Guyett in *Sacred Plant Initiations* says of St. John's Wort: "We know that all plants carry Light, yet St. John's Wort is a supreme case in point. The fact that this plant is bringing *so* much Light to humans is of massive significance to our healing and to the evolution of human consciousness. St. John's Wort is literally leading us to *enlight*enment."

I have often wondered what it means to be enlightened, and I recall the image of Jesus with a halo around his head. Does his halo signify that he was so filled with light that others could see his light? And if we all become filled with this much light, will we emanate a visually perceptible light in our aura? How, then, do we become this light filled? Perhaps the answer lies with St. John's Wort.

Healthy Boundaries

St. John's Wort has been known through time as a plant that offers major protection. Its Latin genus name, *Hypericum*, comes from the Greek *hyper eikon* and gives us indication of this: *hyper* means "over or above," and *eikon* means "image, icon, or apparition." I see that this attribute of protection may contain multiple meanings in its origins. One interpretation of *hyper eikon* is that one is being protected from evil spirits (apparitions). Another interpretation is that one is being protected by an icon or holy image. Perhaps it was the light emitted by the icon that brought protection or a sense of protection. When St. John's Wort was hung over the image of Jesus, Mary, John the Baptist, or angelic beings, protection was guaranteed. These icons could also have been avatars, such as Buddha, Kuan Yin, or Ram Dass, or representations of the divine, such as the Great Spirit, Cosmic Consciousness, or All That Is.

One aspect of protection is maintaining healthy boundaries, which means not allowing any energy, whether human or nonhuman, to enter into your energy field unless invited. When an uninvited energy enters your field, this is considered an intrusive energy. One of the plant signatures of the *perforatum* species is called such because their leaves contain tiny perforations or holes and when held up to the sun allow light to shine through. When there are holes in your energy field, intrusive energy has free access to wander in and out. It's like an open-door policy, or what I like to call "Swiss cheese aura." St. John's Wort has the ability to fill these holes with light and protection from intrusive energy.

Another challenge to maintaining healthy boundaries is when a person is so attracted to your brightly shining light that they want to take light from you instead of generating light from within themselves.

This envy can become quite harmful and draining to your energetic field if allowed to go unchecked. When you feel this crossing of your boundaries happening, call upon St. John's Wort to serve as a mirror, reflecting the person's own light back to them, so they realize it is not necessary to take light from another.

Dispelling Depression

St. John's Wort is famous as an herbal antidepressant because it acts much like pharmaceutical selective serotonin reuptake inhibitors (SSRIs). The National Health Service (NHS) in the United Kingdom explains, on its website, how SSRIs work.

> It's thought that SSRIs work by increasing serotonin levels in the brain.
>
> Serotonin is a neurotransmitter (a messenger chemical that carries signals between nerve cells in the brain). It's thought to have a good influence on mood, emotion and sleep.
>
> After carrying a message, serotonin is usually reabsorbed by the nerve cells (known as "reuptake"). SSRI's work by blocking ("inhibiting") reuptake, meaning more serotonin is available to pass further messages between nearby nerve cells. . . .
>
> Common side effects of SSRIs can include:
>
> - feeling agitated, shaky or anxious
> - diarrhoea and feeling or being sick
> - dizziness
> - blurred vision
> - loss of libido (reduced sex drive)

St. John's Wort, on the other hand, does not have these negative side effects. (***Note:*** If you are taking an SSRI medication, choose a St. John's Wort essence to ingest during your initiation since St. John's Wort should not be taken together with SSRIs.)

The spirit of St. John's Wort also helps to relieve depression because it fills one's energy body full of light and counters spiritual malnourishment (when one's spirit is not being adequately fed). I dream of a spiritual wasteland.

I wander aimlessly through a vast desert of scrub and rock where there is seemingly very little life. What is this place that is so empty, lacking anything that feels good? I wander for what feels like days with no water, food, or encounter with other life-forms. How do I get out of this desolate barren place? Even though I want out of this place, I feel like I must go on, something is pulling me. I notice up ahead a glimmer of light, which I am now desperate to reach. As I approach, I see a green shoot emerging. I fall down on my knees and begin to weep. As my tears water this burgeoning life, I sense a spark inside me. Is this my internal spirit fire reigniting?

This dream came at a time in my life when I was experiencing deep depression and I felt like I couldn't access my joie de vivre. It jolted me out of my amnesia and reminded me that I could receive help from the green beings. The ally that I went to was St. John's Wort, and over time, my spirit flame was rekindled and my love for life returned. We all have times when life circumstances consume our vitality and our own inner strength wavers. Perhaps these times are necessary to show us that there *is* light at the end of the tunnel, and this light reflects our true essential nature so we can see, once again, that we are here to walk a path of beauty.

SIXTEEN
Rose

Rosa rugosa

The second plant that stepped forward during the time of the COVID-19 pandemic, when the world was closed to most in-person contact, was Rose, who agreed to be the elder initiator, virtually, so folks could begin to heal their broken hearts from the devastation that came in the form of a virus, which also created equally devastating responses from many humans. Rose began long before the initiation ceremony to prepare the initiates by showing up in a most personal, unexpected way.

The Softness of Rose

Martha, a former student and dear friend, surprised me with a gift that arrived in the mail. Martha is a gifted weaver, and I had received special weavings from her in the past, but this time I was stunned by the gift she blessed me with. I excitedly opened the package to find a ball of what seemed like yarn except that it was exceptionally soft—so soft I couldn't stop rubbing my face with it. This unprecedented soft fiber was like nothing I had ever encountered. When I was able to talk with Martha, she shared that this was Rose fiber made from the stalks. I was stunned as I had no idea such a sublime substance existed. I had felt the fluff of dandelion and the silky tufts of milkweed, but this fiber took softness to an entirely different dimension.

While talking about the Rose fiber with Martha, I discovered that she didn't know I was facilitating a Rose initiation online, but of course, Rose had already stepped forward to be the initiator, and this was her way of bringing her softness, literally, to all the initiates. I asked Martha if she would send me a bit more of the spun fiber, and she did. Then I cut long lengths of the spun Rose fiber for each of the initiates and sent them in the mail in time to arrive prior to the initiation. I suggested that each person wear the Rose fiber on their bodies, in the form of a necklace or a wrapped wrist band, and to let their hearts begin to experience the beauty, love, and healing from Rose in the form of softness.

During the very first initiation I did with Rose, eight years before this online initiation, the felt sensation that I received was softness. My skin was soft, my clothes were soft, even the air around me was soft—everything was soft. I was in a dream of soothing ethereal delicacy where I was held in exquisite tenderness. Throughout the initiation, I realized that the healing of my heart began with this softness. For so many years, I had carried a hard edge where I closed off a part of my heart that had been so damaged. Finally, in the loving embrace of Rose, I could relax and let the softness of Rose massage my rigid heart.

Healing the Mother Wound

I remember that during the first initiation with Rose, the particular bush I gathered petals from for the elixir was given to me by my daughter. I love this bush so much partly because I love her so much. As I gathered Rose petals for creating the elixir infusion, my heart wandered to my relationship with my own mother. I felt a constriction in my heart and knew there was work to be done on the mother wound in my heart. The opportunity came when I engaged in a Greenbreath journey with Rose.

Breathing deeply, I tuned in to my mother when the ancestral song began to play. I was being birthed, and once out of my mother's watery womb, I emerged into a glaring, sharply shocking, and seemingly violent world in which I was immediately whisked away to be poked, prodded, and weighed. If that wasn't enough, stinging drops were put in my eyes. What is going on? When do I get to smell, touch, and taste my mother? Will I be able to see her face with these drops in my eyes? I long to hear her cooing voice. It was hours before I met my mother, and she was so tired she barely could look at me, let alone hold me. I was taken away again and placed in a strange bed where I cried myself to sleep. This was my welcome to this new world. As I continued to breathe, I was stunned by the scenario that was unfolding. Was I actually remembering my birth? A shift occurred in the music, and I felt Rose wrap me in her softness, and I began to cry. I implored her, "Please Rose, heal this wound of lovesickness."

Bonding with my mother never happened on the very first day of my life, and this continued throughout our relationship. I always wanted my mother's love, but felt she loved my two other sisters more than me. This created a longing that I carried into my adult life where I felt others (mostly women) were loved more than me. Even though this yearning to be loved by others was not rational (nor based in reality), it certainly was felt on an emotional level because of the imprint left by the lack of bonding with my mother and my perception of the small amount of love I received from her compared to my sisters. Over the years, I have grown into my own self-love with the help of my plant allies and by engaging with that which nurtures me and surrounds me with good vibes. I began to forgive my mother when I learned more about the ancestral burden of lack of maternal bonding handed down through my mother's maternal line, and the forgiveness is now complete as I laid that ancestral burden down by deeply bonding with my daughter during her birth and early years.

The Intimacy of Rose

There are so many ways to experience intimacy, and Rose is one of the premier teachers in this department. She engenders the very essence of intimacy, which is deep connection. On a physical level, the skin-to-skin contact with her soft petals stimulates the imagination of lying in a bed of Roses, experiencing the sensuousness of both her alluring fragrance and her luscious petals, much like Cleopatra who charmed Marc Antony to lie with her in a room full of Rose petals.

Intimacy spills over into the emotional arena when sharing true feelings and thoughts with the other. When I sit with Rose I find myself bubbling over with my sequestered emotions, knowing that Rose will always be honest in her response and nonjudgmental in her guidance. I can trust Rose to see me, hear me, and understand my vulnerability while wrapping me in safety. This level of intimacy can be hoped for within the human milieu but usually ends up only being longed for.

Spiritual intimacy comes so easily with Rose as I find myself praying with Rose on a regular basis. There is such heartbreak on all levels, it seems like I'm praying for the very heart of the world to be mended, or at least to suffer less and not ache quite so much. Of course, this prayer is for myself, as well, when I feel my heart so filled with grief for the tremendous loss of so much beauty on this planet. Computer-generated estimates are that up to 150 species of plants and animals go extinct *every day*. If we are to know and experience Nature as our kin, it's unfathomable to grasp that an entire family of relations is wiped out in one day. And yet, here we are, on the brink of the sixth extinction spasm. I have to ask myself and Rose if there is another way to look at this. There is always a gift in Nature, but sometimes our lack of vision keeps us from seeing it. Visionary and New Earth advocate Dr. Zach Bush sees the beauty on the other side of extinction, as he explains during a podcast with Rich Roll:

> We are now in our death cycle, which means there's never been so much opportunity for rebirth and so this is where we're at right now.

We have one moment to rebirth and in that rebirth we can move into this creation mode where we're no longer stuck in the victim/perpetrator mentality. What is necessary is that this leap is alongside the concept of forgiveness in this movement from the belief of scarcity into the state of abundance. So if we were to do this move from scarcity to a state of realization of abundance as this earth moves into more abundant life every iteration, after every extinction there is more life, more diversity, and more intelligence every time and it's stunningly beautiful. We go from palms and ferns before the last extinction 55 million years ago to deciduous trees and wildflowers everywhere. That necessitated an extinction to get that level of new viruses into the atmosphere to code for the new life. Viruses are not living things; viruses are the packets of genomic information that are the possibility of the future. Those species under extinction-level stress are putting out new genetic information that we call viruses or exosomes and they are literally releasing the next iterations of those species that could come next. As the dust settles and this death turns into a rebirth, which happens again and again on this planet, the rebirth is going to be beautiful. We just have a few years to decide whether this new humanity is going to be part of that new future. Are we going to rebirth with the planet or are we going to be part of this sixth extinction event? Are we the cataclysmic event that allows for all that beauty and explosion of Nature to happen over the next million years? Is our gift to the planet that we are the asteroid this time, that we are the thing that destroyed the earth so that it could be rebirthed into a higher level of intelligence and beauty? Maybe that's not that bad of a story in the end.

Rose reminds me of up-close encounters together and how intimate they can be. I remember being at a Rose initiation in California years ago, and we were at the end of our time together. As I was offering my gift to Rose and speaking as eloquently as I possibly could, I began to weep. My affection for Rose spilled out of my eyes in the form of tears, and I realized that each of these salty pearls were precious drops of love. I started to wipe them away when the Rose in front of me started to quiver. I stopped and in that moment knew that Rose didn't want to waste these gems. I put each tear on my finger and anointed the leaves and petals of Rose, sharing this intimate moment with my beloved.

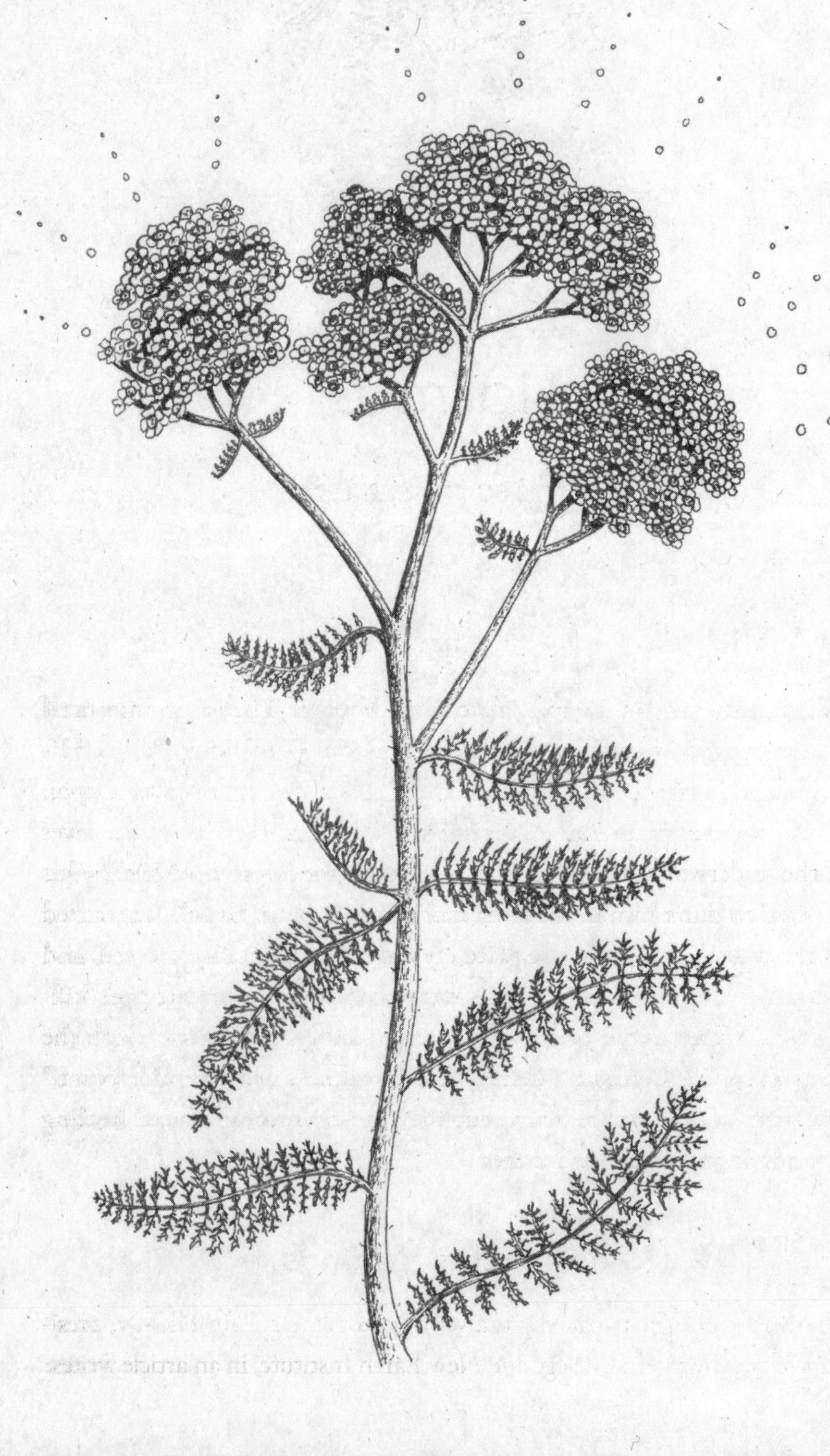

SEVENTEEN
Yarrow

Achillea millefolium

The namesake of Yarrow is Achilles, born of Thetis, an immortal sea nymph of goddess stature, and a mortal Greek king, Peleus. His mother is heartbroken to think of outliving her son, so she takes it upon herself to attempt to make him immortal by dipping him in the river of the underworld, the River Styx. However, she has to hold him by his left heel to dunk him in the river, and this part of his body is untouched by the water. This is the one place on his body that is unprotected, and eventually the Trojan prince Paris wounds Achilles in his left heel, killing him. Here lies the origin of the expression Achilles' heel, which the *Oxford English Dictionary* defines as a "weakness or vulnerable point." The gifts of Yarrow are many, but they mostly revolve around healing wounds of one variety or another.

Patriarchal Wounds

The wounds of the patriarchy is a very big topic. Dr. Ann Filemyr, president of Southwestern College and New Earth Institute, in an article writes:

> Patriarchy defines a strictly binary code for masculinity and femininity and is intolerant of multiple or fluid/trans gender identities. This gender imbalance is perpetrated through religious ideologies, colonialism, post-colonialism, legal and economic systems, cultural norms and family patterns. The patriarchal wound may be experienced and expressed differently by boys, girls, men, women, trans persons, lesbians, gay men and other gender and sexual groups. However, all persons suffer from its impact.

Generally speaking, in a patriarchal system the feminine, which has to do with receptivity, feeling, and heart, is considered inferior to the masculine, which is perceived as power, action, and mental acuity. Of course, there are men who display their feminine side, but this is mostly frowned upon by both men and women, and likewise, women who display strength and a can-do demeanor are not viewed in a positive light.

Yarrow can bring healing to this massive wounding. My student Valerie shares how she has been helped by Yarrow to heal the wounds of patriarchy.

> *I grew up in an extremely patriarchal household. Though I understood this "going with the program" of patriarchy was a trauma reaction brought on by generations of unhealed sorrow and pain. I am the first of my lineage to not experience direct separation from land, community, war, and starvation; nonetheless, I inherited a tremendously complicated relationship to my father and patriarchal violence. My initiation with Yarrow was a turning point in healing what felt frozen in time. During Greenbreath with Yarrow, I was able to see some of the roots of the pain of my male-identified ancestors who went through tremendous suffering as they witnessed their partners and children wounded by both the church and state. I saw how the horror they experienced crystallized as rage and conflict between male- and female-identified partners, generation after generation. I saw Yarrow able to grow through the cracks in this system of patriarchy. How they are a plant spirit that can help ease the burdens we carry (no matter what our gender). Yarrow helped me to bear witness to*

the pain of the wounded masculine—without taking it on as my own—and from that place of witnessing I begin a journey of somatic healing around my own history of suffering sexist violence. I begin to have clear relationships with male-identified folks after this.

The day after my Yarrow initiation, I witnessed a tremendous act of violence between men on the street. My grandmother also began to share her story with me about the violence she encountered in her own life during war and immigration. I could see how that impacted how she raised my father and then how he raised his children.

None of these stories are simple. So much of the violence I have witnessed has to do with class, with resource distribution, with the lack of choices people feel they have under modernity. I began to realize how patriarchy is upheld by other systems of oppression, and Yarrow became my ally in dismantling (as bell hooks calls it) "imperialist, capitalist, white supremacist patriarchy."

I feel like my initiation with Yarrow helped me to understand that my feminism could be a much more powerful tool when rooted in liberation for all beings. Yarrow helped me to see how masculine-identified folks are also wounded by patriarchy, and at the root of holistic and ecological feminism, we must be allies for each other's healing. Yarrow is my guide in having healthy boundaries and regulating my sense of what I can handle so I don't move into codependent tendencies but instead allow my relationship with myself and people of all genders to be interdependent. With Yarrow by my side, I can express the full range of my emotions and hold space for those in my life to do the same. I know what's mine to hold and feel protected by the warrior spirit of Yarrow.

Filemyr further notes in her article that "the limiting of human experience and expression through overt and covert forms of gender-based violence distorts and undermines our full capacity as members of humanity." This violence occurs in many ways for both men and women. For men the patriarchy has damaged young men (boys, almost) by sending them to war against their will. During my youth my male friends were being drafted to fight a war that they vehemently disagreed with and in the end was shown to be one of the biggest mistakes of the American modern era. Also men who display feminine tendencies are targeted and oftentimes violently abused by other men because the patriarchy says this is unmanly behavior. However, women have been violently abused emotionally, mentally, and physically for eons as a result of patriarchal beliefs that are so deeply embedded that men feel they have a right to abuse women. The good news is Yarrow is here to help men and women alike.

Filemyr also reminds us in this reflection of the damage the patriarchy has perpetrated on Mother Earth:

> The patriarchal wound has profoundly ruptured our sense of belonging to the Earth. The patriarchal paradigm also insists that nature be reduced to *resource* instead of understood as it is: *life source.* As a result of this error in thinking and subsequent erroneous economic policies and practices that are centered on private wealth accumulation instead of mutual flourishing, we are facing mass extinction, the death of the seas, anxiety, isolation, despair and suicide.

Yarrow is stepping forward through the initiation process to help us heal our disconnect from Nature. As we enter more fully into co-creative partnership, the notion that we harm Nature is unthinkable. Earth is our mother, but Nature is our grandmother, and no one (who has more than two peas in their pod) harms their grammy.

Balancing Masculine and Feminine

We experience both the masculine and the feminine in our beings. The two aspects manifest on the right and left sides of the body—masculine is on the right side and feminine is on the left side. When there is an imbalance between these two aspects, we ask the spirit of Yarrow to shift the balance by working together with Mugwort to transfer the energy. Mugwort is an excellent mover of energy, and when working in tandem with Yarrow, balance can be achieved. I received this movement meditation from Yarrow as a way to balance the masculine and feminine:

In a relaxed stance raise your hands, with the right hand facing out and the left hand facing in. Push out with the right, and pull in with the left. Let your masculine send out to the world, and let your feminine take in the world. Begin to move in a flow of sending and receiving. Let your imagination flow in and out, with your feminine and masculine in sync with equal movement. Now, add your breath to the movement: breath into the feminine and breath out with the masculine. Equal inhalation as exhalation, as the masculine and feminine move in a coordinated dance. In this syncopation of breath and hands, where one depends on another, an equilibrium is attained.

Cut to the Bone

Herbalist and friend Matt Wood refers to Yarrow as the "cut to the bone" plant, suggesting that it helps to heal wounds that go deep, which is in keeping with what we have already seen in the previous stories. Years ago, when my daughter was in college, she worked for a local pub and eatery as a vegetable chopper (sous chef is a bit of an elevated title for this particular establishment). Her cutting board was made of heavy stone, and one evening it slipped off the table and the edge landed on her foot, cutting it badly as well as giving her quite the bruise (it's a miracle her foot wasn't broken). She called me in a panic, wanting to know what to do. I had recently enjoyed wildcrafting the white flowers of Yarrow and then drying them in my warm attic space. I remembered what Matt said about this wild-growing plant and knew immediately this would be the plant to help heal my daughter's foot. But I needed to get it to her as soon as possible, and she was in Burlington, Vermont, while I was in southern Vermont, at least a two hour and fifteen minute drive (when going a wee bit over the speed limit). I put the kettle on to boil water and prepared a quart-sized infusion before I left. It would be perfectly steeped by the time I arrived in Burlington. By the time I arrived, she had managed to hobble back to her apartment and was sitting with her foot propped up with an ice pack on it. I poured the infusion into a bowl, and she started soaking her foot. I was not able to stay with her that night as I had appointments in the morning that I couldn't change at this short notice. I instructed her to soak it as long as she could that night and then, when she went to bed, to wrap a yarrow-soaked cloth around her foot, saving the rest of the liquid to soak again the next day. I also told her to stay off her foot. She then informed me she absolutely had to work the next day. I encouraged her to at least sit with her foot propped up while she chopped vegetables. The next evening, when I was able to talk with her, she said the cut had completely closed up, the swelling had gone down, the bruise was already fading, and she was no longer in pain—all of this in less than twenty-four hours. I was impressed.

I had recently been working with a client who had some energetic cords with an old boyfriend that were not healthy. It seemed important to cut these cords, but I had never done this type of work with a plant spirit before. I began to wonder if Yarrow might be the one for this job. Could Yarrow actually *do* the cutting on an energetic level that would bring about healing? I went and sat with Yarrow and entered into a dream journey with the plant to learn more about Yarrow's abilities in cord cutting. It turns out this is one of Yarrow's gifts, and I was shown how to facilitate the cord cutting, with Yarrow performing the actual cutting.

Several years later, during an initiation with Yarrow, we were engaging in a dismemberment journey, which is when the old you is replaced by a new you. That which no longer serves you is removed, and what does serve you fills the space of you. Gina was struggling in her journey, and at one point, she began making chopping motions. Later, when we were in circle and she had an opportunity to share about her journey, she said that there were so many huge energetic cords that looked like tree trunks that she had to help Yarrow chop them off before the dismemberment could even begin. I was amazed at this but so grateful to Yarrow for sharing this gift and helping Gina to understand how to eliminate these energy-sucking cords.

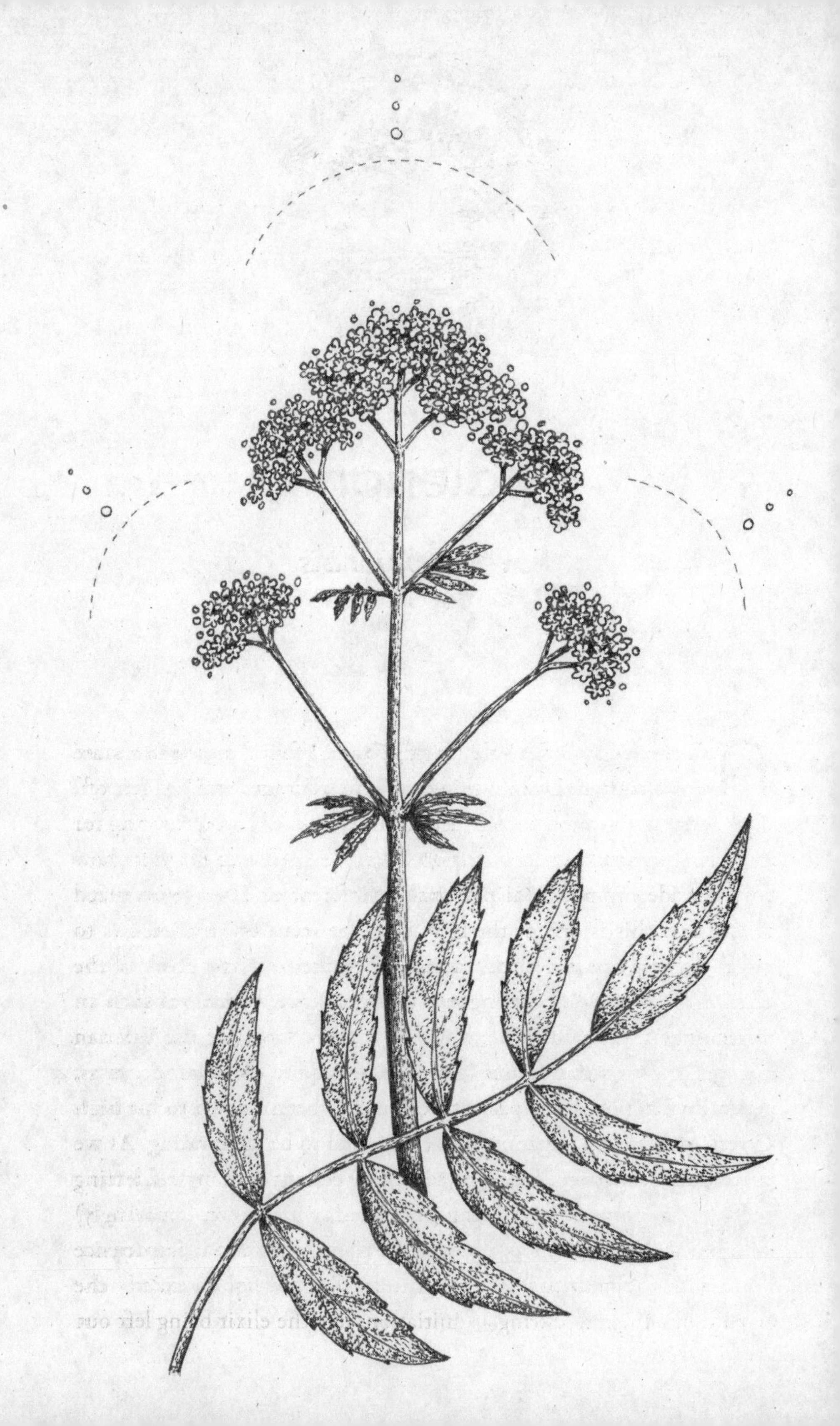

EIGHTEEN
Valerian

Valeriana officinalis

As I entered into the Valerian initiation, I found myself in a state of distraction. I forgot things, misplaced items, and just felt off. This is out of character for me, especially whenever I am preparing for a teaching situation or an initiation. I have learned over the years how to put aside my potential personal interferences. Then it occurred to me that this is one of the gifts that the spirit of Valerian has to offer—to help one not be distracted. With the very first drink of the elixir, I found myself entering into calm patience, which was such an intriguing use of words. The calm part I understood, because Valerian is a well-known sedative, but "patience" was quite the conundrum as, generally, I'm not a very patient person. I've been known to hit high C very quickly in a situation that I may find to be aggravating. As we entered the initiation I found myself being patient with myself, letting go of my annoyance over being distracted, which then (amazingly) eliminated the distractions all together. I also let go of any impatience I had with the multitude of little things that did not go exactly the way I'd like them to during an initiation, like the elixir being left out on the counter instead of being refrigerated, or the smudge stick not staying lit, or a thunderstorm breaking out the minute we are ready to go outside and sit with the plant. Then I remembered that this initiation is in the hands of Valerian who is running the show as the plant elder and the initiator. I am merely the facilitator. So I relaxed as I stepped into receiving the gifts Valerian was offering all of the initiates and myself.

The Hypnotic Nature of Valerian

I have included Valerian in an herbal sleep formula that I make so I know it well as a sleep aid. However, during the initiation I discovered that it is somewhat psychotropic, or maybe I should say mind altering. I noticed that my visual and auditory senses were more acute. The wind in the trees was louder and more accentuated. It was almost as if the trees were telling their story via the caress of the wind. The dew on the grass in the morning was refracting the light, making tiny rainbows everywhere. I have seen this before, but now the light sparkled more brightly, and there was a sound accompanying the rainbows that was like the clear tinkling of bells. I marveled at these events and was awestruck by the exquisite beauty of this jewel of a planet we call home. I realized that Valerian was not altering the brain chemistry like a true psychotropic plant does, but instead its action of calming of the nervous system was allowing me to slow down and stop the mind chatter enough so that I could be in the Now moment, when it is possible to step through the portal into other dimensional realities. The experience I was having with Valerian was that of accessing synesthesia, which is the use of one sense in the capacity of another. In other words, I was hearing what I was seeing and I was hearing what I was being touched by. This has rarely happened to me, so I was thrilled when I discovered that Valerian carried this gift and chose to share it with me.

Bridging the Worlds

One summer I was visiting my friend Carole Guyett, who is the originator of plant initiations, and she was planning a Valerian initiation. We talked about what it might be like to bridge our two sanctuaries and consciously connect our collective energies, intending to nourish the web of life. During my visit we harvested Valerian blossoms together with the intention of adding them into each of our respective elixirs. Once I returned home I likewise harvested Valerian flowers from my land, which I sent to Carole to make an infusion to go in her elixir, so the two elixirs had blossoms from both of our lands of Sweetwater Sanctuary in Vermont and Derrynagittah in County Clare, Ireland. Our intention was to strengthen bonds of peace and harmony between all lands with the blessings of Valerian spreading across our beloved planet.

When we were walking the labyrinth during the initiation, I had a vision of Gaia, the Earth goddess, with one foot in my labyrinth and one foot in her labyrinth, creating a rainbow bridge between the two locations. We could access each other's experiences, joined via the bridge of Gaia, connecting in the unified field of Valerian. The primary gifts that Valerian shared with the Ireland group of initiates echoed the experiences we had in Vermont. Carole summarizes their initiation.

> *Valerian graced us with* deep peace, *bringing stillness and simplicity where gentle restoration was present. Valerian took us into the darkness, where deep transformation and relief from suffering was possible. We experienced death and rebirth, dying to the old and being reborn to the new. Bast, an Egyptian goddess carrying the energy of a lioness (and later a cat), visited us as well as other black-colored animals (bear, panther, and raven). We entered the dreamtime as conscious dreamers intentionally dreaming the world we have visioned into existence. As we opened to the possibilities of a New Earth, many blocks were removed in order to achieve this. Our creativity was stimulated, and previously unseen potential began to flourish. We were blessed by the many gifts of Valerian, and we emerged as more intact seekers of a new paradigm.*

Rededicating to Partner Earth

It has been over a quarter of a century since I wrote my first book titled *Partner Earth.* I had been involved in the herbal world for about a decade at that point, but my sense of the importance of changing our relationship to Earth was in the forefront of my consciousness. I decided to write about my experiences and ideas even though I knew nothing about how to write or the process of publishing a book. I look back on this piece of writing, and I cringe a bit in my elementary approach to writing, but at least it was an attempt to convey something that all these years later has become paramount in the face of what is being called a climate crisis.

The concept of *Partner Earth* illuminates the imperative that we shift our adolescent way of relating to Earth and evolve into a mature co-creative partnership where we are consciously engaged with the source of our sustenance—a living Earth that encompasses all of Nature. These were my words from *Partner Earth:*

> We are no longer a child to Earth. We are now becoming an equal co-creative partner. Earth is tired of being our mother—our adolescent period has worn her thin. She, too, is ready to evolve beyond the parent-child relationship. She is crying for a true partner, and so we answer her in wholeheartedness. She now becomes Partner Earth instead of Mother Earth. This may seem blasphemous to some, most notably the indigenous peoples who have never abused their mother. The Earth has always been our mother; how dare I suggest anything else. One friend said, "We'll have to change all our songs." We don't need to change the songs that came before; we can create new ones. I intend no disrespect. I merely long for us all to mend the hoop, to enter the full family circle again. Changing our relationship doesn't mean obliterating what came before—quite the contrary. It's important to embrace our childhood experience and learn from it instead of denying it. Earth will always be our mother but we can evolve into an expanded way of relating. Our relationship matures into a co-creative partnership.

This vision continues to be my primary focus, and Valerian reminds me of its importance. While walking the labyrinth Valerian says, "Rededicate yourself to Partner Earth by remembering what you were told years ago. Bring your teachings of Partner Earth to as many people as you can. Write again about this but in a new way." This came in 2019, and as I sit here writing to you four years have passed, but here I am. Sweet Lady Gaia, I do not forget you, I do not forsake you, I hear you, I love you.

NINETEEN
White Pine

Pinus strobus

White Pine became an ally of mine soon after I moved to Sweetwater Sanctuary. As I recounted in chapter 10, I had a most profound and intimate experience with White Pine during a transformational breathwork session years ago. That ecstatic session was the beginning of my love affair with White Pine, which continues to this day.

White Pine serves as the western gate of Sweetwater Sanctuary where one can experience deep peace beneath her boughs and where she opens a portal for the ancestors to enter. Here, her protection and incredible strength is felt. When I tap into the felt sensation of White Pine, I experience her "tower of strength," and my posture changes to standing tall with my shoulders upright and balanced in the center of my being. Receiving the incredible gift of standing in strength has made me feel invincible at times and has supported me through some very challenging times, such as during my journey to Ireland after a volcanic eruption, which I described earlier in this book.

Tree of Peace

White Pine has long been associated with peace, as this article from the Indigenous Values Initiative describes.

> Over 1,000 years ago, the Five Nations were brought together in peace at Onondaga Lake by the Peacemaker and Hiawatha (Hayenhwátha'). Together they planted the Great Tree of Peace (Skaęhetsiˀkona) and created the Haudenosaunee Confederacy. This is where Skä·noñh began anew.
>
> The Tree of Peace is a metaphor for how peace can grow if it is nurtured. Like a tall tree, peace can provide protection and comfort. Like a pine tree, peace spreads its protective branches to create a place of peace where we can gather and renew ourselves. Like the White Pine, peace also creates large white roots (tsyoktehækęætaˀkona) that rise out of the ground so people can trace their journey to the source.
>
> If anyone truly desired peace they could follow the sacred white roots of peace to the capital of the Confederacy, here at Onondaga, where they would learn of the words of the Peacemaker. His message is that we all can nurture the "Tree of Peace."
>
> The Peacemaker had the warriors uproot a great white pine under which left a gaping hole. The 50 chiefs and warriors threw their weapons of war under the Great Tree where an underground stream carried the weapons away and it was lifted back upright. This is the origin of the phrase "bury the hatchet." The Peacemaker said that the Chiefs will be standing on the earth like trees, deeply rooted in the land, with strong trunks, all the same height (having equal authority) in front of their people, to protect them, with the power of the Good Mind—not physical force. On top of the tree sits an eagle who serves as an ever vigilante [sic] protector of the Peace.

My student Suneet shares her insight during the initiation about the Tree of Peace: "The White Pine Tree, also known as the Tree of Peace, gifted me so many teachings over those few days, that even now, years later I am still learning from them and recommitting to them. White Pine taught me the meaning of being an ambassador of Peace, something much needed in my family still to this day, both immediate and extended" (see color plate 15). And her song of White Pine speaks of peace:

Hear our thoughts
Hear our prayers
Peace Be Upon Us
Standing tall
For us all
Peace Brings Harmony

Mark describes the message of peace that White Pine shared with him during the initiation: "Peace is not something to attain as much as something to maintain, not a thing to acquire or a condition to establish, but instead a willingness to be what and who we are in relation to everything else in the world, each according to their nature."

Supporting Diversity

Growing up, I was always the one in my family or group of friends who chose the path less traveled. I was a bit different and had what some would call far-out ideas. This was often a lonely place to find myself, but it was me. This is partly why I was always so comfortable in Nature. Nature did not negatively judge me. I could just be me, and even more so, I was supported by Nature to be my unique self. It was as if Nature wanted me to be different. Reflecting back on my earlier years, I now see that Nature's encouragement to be one's own unique self is part of the innate diversity that is inherent within Nature. It is diversity that gives life its dynamic quality leading to a thriving interdependent community whether it is human or nonhuman.

After participating in the White Pine initiation, Mark describes White Pine's unique diverse nature as a nonconformist.

You are not the biggest conformist in the world, as many of your limbs grow in whatever direction and length you see fit. Sometimes many of your arms spread out only on your left side or only in one direction on some portion from your torso. Some of your branches droop while some reach upward. Sometimes you grow full and flush, thick in form, while others in their years and years of growing upward have sent out no limbs at all for long stretches of time before deciding to reach out into one unencumbered space and respond to the counterbalanced need of your vertical aspirations.

While Creeping Thyme, Ground Ivy, and Princess Pine hold fast and merge themselves together in a tight-knit community, you reach for the sky. Each and every one of your five-fingered needle bundles drink from the sun. Thank you for your instruction to stand strong, to be exposed and seen in all your irregularities. My strength is to be like you in these ways. It's OK to be old, a bit ragged at times, to be grandiose, too, and sometimes be incomplete and other times abundant in the fullest of shapes.

In our daily lives there are many examples of how a lack of diversity can be limiting and, in some cases, damaging. One of the main areas where we have seen devastation from the lack of diversity, or what is called monoculture, is in farming. The negative consequences of monoculture farming—the practice of planting only one crop over many acres—have been well documented by such organizations as the National Institute of Food and Agriculture, which is a division of the USDA. Slowly, the USDA is beginning to change its views on monocropping, especially since it is not proving as cost effective as it was originally thought to be.

The alternative to monoculture is what is called permaculture, the practice of an agricultural system that is permanently sustainable. The inception of modern-day permaculture came in 1978 when Bill Mollison and David Holmgren, both of the University of Tasmania in Australia, coined the word when they released their book *Permaculture One*. Valentin Kunze, in his article "A Brief History of Permaculture," states:

> In *Permaculture One,* Mollison and Holmgren devise a framework for an alternative, sustainable agricultural system that is based on combining perennial trees, shrubs, herbs, fungi, and root systems in such a way that all elements support and benefit each other. As the framework mimics the functioning of our planet's natural ecosystems, it is meant to be an agricultural system that can be sustained indefinitely—a permanent agricultural system, a permaculture.

Kunze goes on to say:

> Permaculture has grown to be a globally known concept and an immense amount of knowledge, methods and tools have been developed through the contribution of innumerable permaculturists. Permaculture has developed into a holistic life-philosophy that is much more than merely a set of alternative farming practices. Nevertheless, much of *the* focus in permaculture still remains on the *earthcare* ethic, although it has transpired that it is only in connection with a healthy people, a healthy society and economic security that cycles can truly be closed. Rather than simply meaning *permanent agriculture,* permaculture today already stands for *permanent culture,* and this is where I believe the true strength of the permaculture philosophy lies: permaculture was born with the aim of sustainably securing the basis for human survival, but has grown to truly encompass the equally important determinants of *peoplecare* and *fairshare.*

Hundreds of years prior to the rise of modern permaculture, the Maya were engaged in a similar practice through forest farming and milpas in Central America. Charles Mann in his book *1491: New Revelations of the Americas Before Columbus* writes:

> A *milpa* is a field, usually but not always recently cleared, in which farmers plant a dozen crops at once, including maize, avocados,

multiple varieties of squash and bean, melon, tomatoes, chilis, sweet potato, jicama (a tuber), amaranth (a grain-like plant), and mucuna (a tropical legume). . . . Milpa crops are nutritionally and environmentally complementary. . . . The *milpa*, in the estimation of H. Garrison Wilkes, a maize researcher at the University of Massachusetts in Boston, "is one of the most successful inventions ever created."

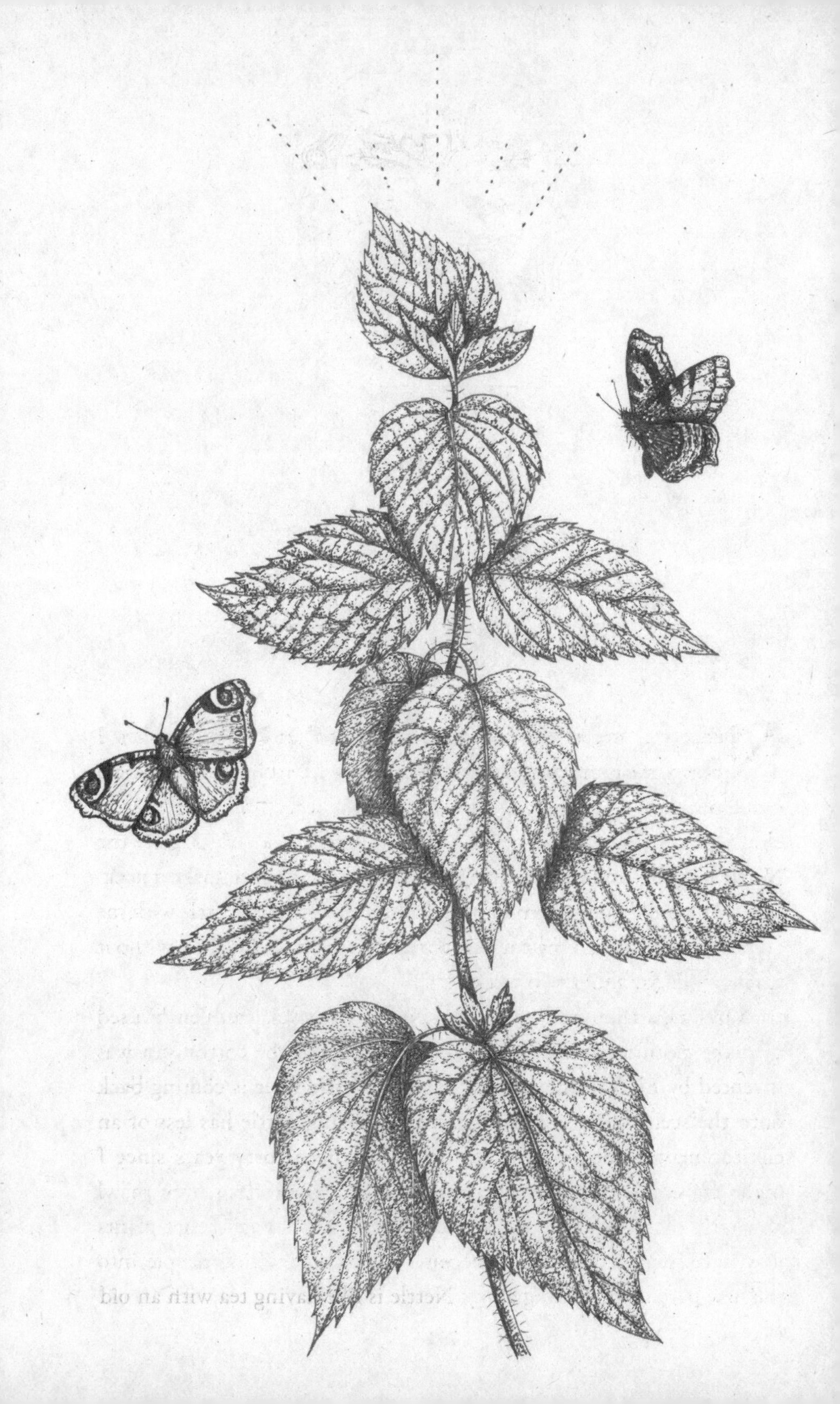

TWENTY

Nettle

Urtica dioica

Nettle was one of the very first herbs I fell in love with when I began my journey with the green beings. I remember making a crude necklace from the plant by rubbing the leafy stalks together to eliminate or break the hairs, which contain formic acid, which gives the Nettle its sting, and then weaving those stalks together to make a necklace that I could wear around my neck. I wanted to have Nettle with me all the time as I began my courtship with the plant, so I wore it without taking it off for about two weeks.

Over two thousand years ago, Nettle fiber was frequently used to make clothing and then fell out of use when the cotton gin was invented by Eli Whitney in 1793. Now, Nettle fiber is coming back onto the scene because growing and processing nettle has less of an environmental impact than cotton. It has been forty years since I made my simple necklace, and now I have a beautifully woven shawl of Nettle that I wear on special occasions. Even though other plants may have stepped forward to become initiators, guiding people into the new paradigm, writing about Nettle is like having tea with an old friend who has always been here and will continue to be here, always having my back.

April, my former student and colleague, shares a different perspective of the fabric of nettle and how it weaves into our lives.

> *In "The Wild Swans," the fairy tale by Hans Christian Andersen, the princess is able to complete all but one of the Nettle shirts before her impending death. She is rescued, but not before clothing the brothers in their shirts. The brother with the unfinished shirt retains a swan's wing instead of his arm, where the sleeve remained unfinished. In my Greenbreath journey, I recalled a past life's sudden ending and my soul's deep disappointment in not being able to finish making a garment for my son. Is it the unfinished things we mourn when we die? Or is it just that before death we regret the not yet completed? Kind Nettle granted me the grandmotherly message that "things are never finished, dear. There's always more weaving to do and more mending. The unfinished or torn can be the opening to heaven." The brother with the swan's wing, I tell myself, could be seen not as a failure but as a symbol of transmutation, becoming spirit, able to go back and forth between the worlds, weaving them together.*

Deeply Healing

Nettle has many nutritive gifts in the form of vitamins and minerals, especially vitamin C and iron, making it a fabulous herb for any chronic conditions. Taking Nettle will increase vitality and well-being and will return homeostasis to one's being relatively quickly. Having a particular affinity for the urinary system, Nettle can remove toxins from the kidneys and restore healthy function. Besides bringing healing on a physical level, Nettle can heal on emotional, mental, and spiritual levels, as well.

On an emotional level, Nettle helps one overcome fears that are unrealistic or overblown. Years ago, I had a student who was afraid of almost everything. When she spoke, she looked like a deer in the headlights, with her eyes wide open as if she had just seen a ghost. Everything we did in class brought up some type of fear; consequently, we spent enormous amounts of time in class talking about those fears. Finally, I suggested she sit with Nettle and ask Nettle to help her with her fears. Of course, she was afraid of being stung by Nettle, which is a realistic fear since folks who encounter Nettle do often get stung. I told her how to approach Nettle and not get stung, which allayed her fear. After working with the spirit of Nettle for several months, she reported that she no longer felt like her life was on hold. For years she had not done certain activities or gone on adventures because she was afraid. Her fear was keeping her from engaging in meaningful, fun, and inspiring activities in her life. Now, with the help of Nettle, she felt free to live her life more fully.

From a different perspective, April describes what at first seemed that it could be fear.

> *I feel in Nettle's presence a place of repose . . . for a time. While sitting with the plant I wrote, "I feel cuddled in her meadow scent, like a mouse in a nest of hay. Where everything is already all right." But then, when I fall into that cozy relaxation, somewhere in the depths of it, my heart rate speeds up, as if under the influence of adrenaline. And I ask myself, what do I fear, in this quiet green moment by a brook? My mind makes up a million answers. Then there comes a new question: Could it be excitement instead of fear? Could some part of me, not yet visible to my mind, be forecasting tremendous joy?*

On a mental level, the flower essence of Nettle seems helpful. Lisa Estabrook, creator of the Soulflower Plant Spirit Oracle deck, describes how Nettle brings about transformation. "Nettle helps you transform foggy thinking so that you can respond to prickly or heated situations by speaking up for yourself, in alignment with your truth, without harboring resentment or attachments that can leave you feeling tired, congested and irritated."

The enlightened eleventh-century Tibetan yogi Milarepa is said to have existed on only Nettle for the years of his meditation in a cave high in the mountains. He is depicted as having a green body, which is said to be from consuming only Nettle. This makes me wonder how the Nettle enhanced his path to enlightenment. Did Nettle not only nourish him physically but also spiritually? Was Milarepa engaged in a massive form of initiation, with Nettle being the initiator? The dieting of plants over long periods of time is a common practice in getting to know a plant deeply. Ingesting a plant for at least forty days is, approximately, the length of time it takes for the plant to move through each level of one's being—physical, emotional, mental, and spiritual. Milarepa then spent months (maybe even years) with Nettle and its formidable spirit, affecting him on his path to enlightenment.

The Ouroboros

While participating in a Nettle initiation in Ireland, the alchemical symbol of the ouroboros came into my awareness. The ouroboros is depicted as a snake or dragon who swallows its own tail. This symbol was adopted in many alchemical circles and represented eternal renewal, wholeness, and infinity. An online article by the BBC explains that the ouroboros symbol "in its original Egyptian context symbolized repetition, renewal, and the eternal cycle of time." It is not surprising that this potent symbol and its meaning was the image that emerged for me while being initiated by Nettle. Earlier, the transformative quality of Nettles was mentioned, which corresponds with the core concept of alchemy—the transformation of base metal into gold.

While in the dreamtime with Nettle, the words *All That Is* kept repeating in my psyche, my phrase for the Great Spirit, God, or Goddess. While investigating the ouroboros, I came across a depiction of the ouroboros on a third-century CE papyrus with the Greek words: "One is All, and by it All, and for it All, and if it does not contain All, then All is Nothing." After reading this, my version of this same concept that continually repeated itself during the Nettle initiation made sense. The magnanimous spirit of Nettle encompasses *everything.*

Transitioning into the New Paradigm

Nettle is one of the plants helping to lead us into the new paradigm and the New Earth. Because of its regenerative and transformative abilities, Nettle is well suited to help us adapt to these rapidly changing times. Because Nettle offers balance, restoration, detoxification, and nutrition, it will help one's physical body to withstand the chemical and biological challenges of these times. In order to rebirth as a part of the New Earth that will flower anew after this spasm of the sixth extinction, we must strengthen our core. Nettle helps strengthen our core by centering our hearts, bringing clarity of mind, and connecting us to unity consciousness. Nettle is a plant ally that will carry us, not just surviving but thriving, into the new paradigm era.

The power from within that emanates from Nettle is not always seen but is certainly felt. We experience this physically when we touch Nettle and receive a sting from the formic acid that is injected into our skin from the hairs on the stem and tips of the leaves of the Nettle plant. On a spirit level, Nettle helps one build one's inner strength. Though this might not be displayed outwardly, a solid inner core is felt by all who encounter a person with such potent energy.

April received a message from Nettle about her future self who has the potential to incarnate during the new paradigm.

> *Part of the experience of walking the labyrinth with Nettle involved contemplation of the question, "What is my power?"*
>
> *Asking myself this, I realized a gift I have, sometimes, of envisioning a good, even glorious future. Through all my lifetimes I slowly evolve, gaining wisdom and sometimes, it seems, losing it to become foolish and baffled again. But evolution does go on, toward something better.*

However, it seems not so much a result of trial and error and error again (and hopefully learning from those mistakes), as it is the power of a future being drawing me forward. And that being isn't me—it's us. Especially when we are woven together, we feel the connectedness, we sense it drawing us forward toward our future beautiful self.

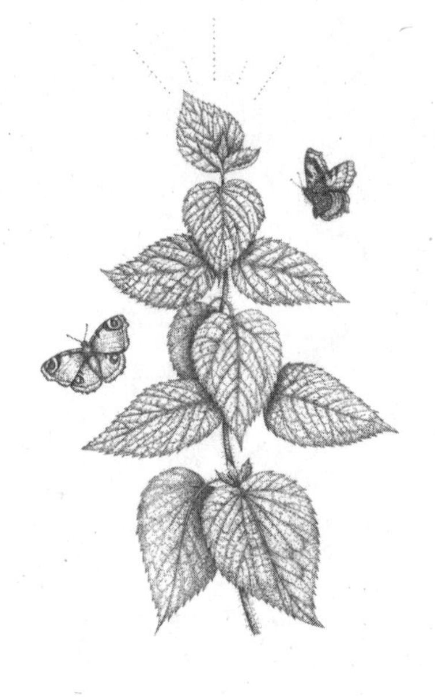

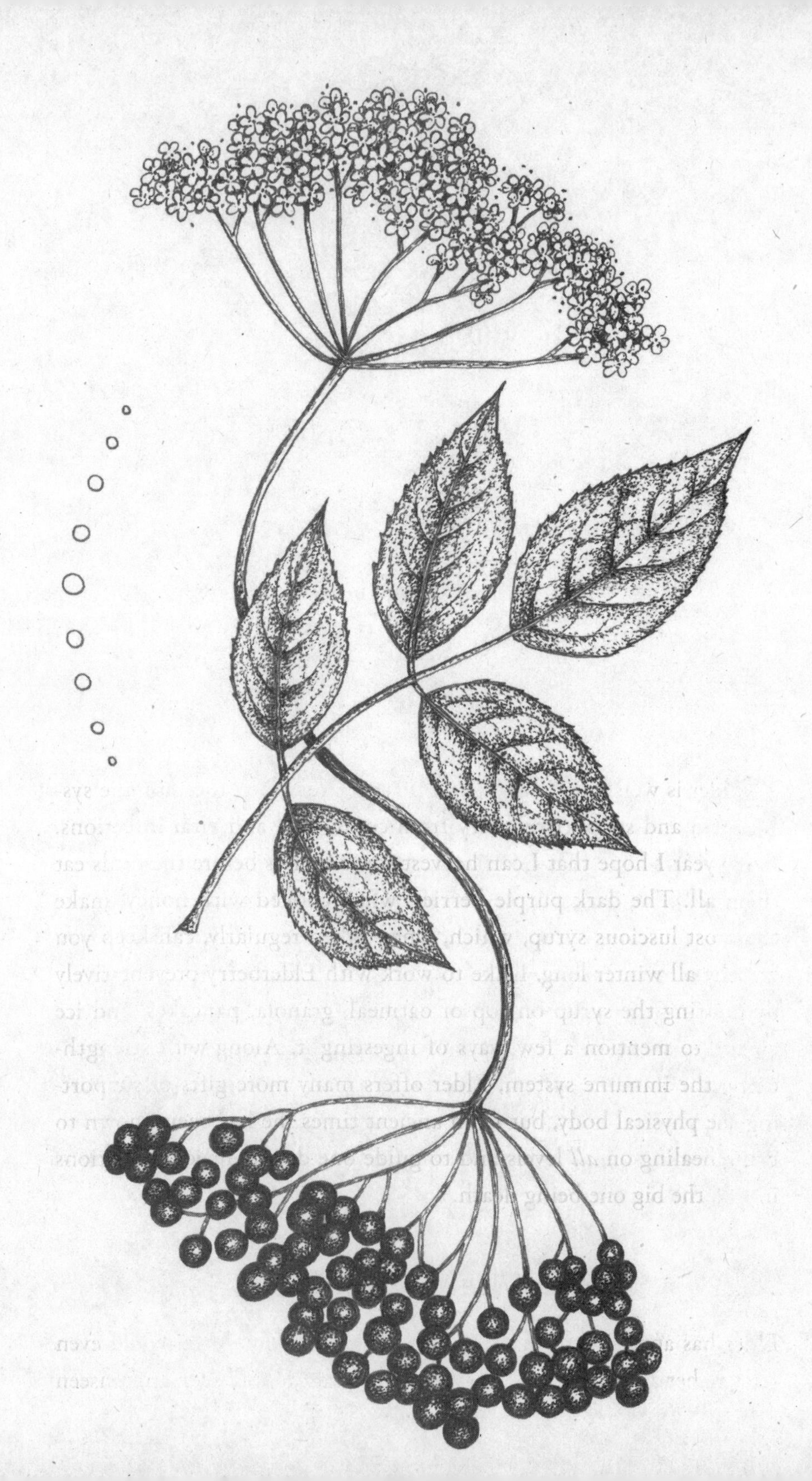

TWENTY-ONE
Elder

Sambucus nigra

Elder is well known for her ability to strengthen the immune system and support recovery from colds, flus, and viral infections. Every year I hope that I can harvest some berries before the birds eat them all. The dark purple berries, when cooked with honey, make the most luscious syrup, which, when taken regularly, can keep you healthy all winter long. I like to work with Elderberry preventatively by pouring the syrup on top of oatmeal, granola, pancakes, and ice cream, to mention a few ways of ingesting it. Along with strengthening the immune system, Elder offers many more gifts of supporting the physical body, but from ancient times she has been known to bring healing on *all* levels and to guide one during major transitions in life, the big one being death.

The Wise Woman Crone

Elder has appeared to many as a wise old crone, or some would even refer to her as a witch, meaning one who heals in both seen and unseen realms. Her very essence and archetypal image is that of the elderly wise healer.

In the Celtic tradition a well-known trinity is that of the Triple Goddess, which essentially corresponds to the three phases of a woman's life—the maiden, the mother, and the crone. However, Lisa Chamberlain, of Wicca Living, broadens this definition:

> [W]hile a woman will proceed linearly through these phases in a literal sense during her lifetime, each aspect of the Triple Goddess has qualities that all of us—male and female—resonate with at various points in our lives. Indeed, the three-fold form of the Goddess could be said to reflect the complexities of the human psyche, as well as the cycles of life and death experienced by all who dwell on Earth. . . . The Crone is the wise elder aspect of the Goddess, and governs aging and endings, death and rebirth, and past lives, as well as transformations, visions, prophecy, and guidance.

Elder embodies the crone aspect of the Triple Goddess.

I recall tales of folks cutting Elder bushes or trees without permission, and how thereafter, life for the cutter did not go well. Many years ago, when I lived on a commercial farm in the Hudson Valley of New York, the long, steep driveway was being regraded. When I discovered this activity was in progress, I, in a panic, rushed to the area just in time to stop the machine operator from destroying a large Elder bush that grew at the edge of the driveway. There was a bit of negotiating between the farmer, machine operator, and myself before an understanding of the potential consequences made clear that it was not a good idea to demolish the Elder bush. I wonder if saving Elder trees and bushes from demise, or at least discussing with them the possibility or probability of their own death when it may be necessary to clear land, could alleviate much unnecessary suffering, if only one could work co-creatively with Elder to communicate the needs of both the person and the plant.

Lucinda describes her encounter with Elder and how she had an inherent knowing about this plant.

I had been walking around the fields and lanes near my home looking to gather elderberries but had had no luck as they had all been eaten by wildlife very early that year. On my way home, I felt a tug in my abdomen and a very clear internal voice saying, "This way!" I turned and saw that there was a hedgerow on the side of a field that I hadn't walked down before. About halfway down, I met the most incredible grandmother Elder tree. She literally shone out and was filled with ripe berries. I knew in my heart not to gather the fruits immediately before really taking time to connect with her and honor the wise and generous being that she was. It was a wonderful meeting, and I understood from it, on a visceral level, why the Elder tree was so revered in folklore as a guardian for humans, wildlife, and the otherworld. That meeting was the beginning of a wonderful friendship.

The Veil between Worlds

In my co-creative partnership with Elder, I have come to know her as one that easily operates beyond the thin veil that separates the two worlds of the seen and unseen realms, the explicit and implicit orders, or the waking reality and the dreamtime. Elder helps humans to operate in this veil space by working through the sixth chakra, known as the third eye, where one can access one's clairvoyant abilities. Carole Guyett, in her book *Sacred Plant Initiations,* says of Elder that "the tree opens up vision, clears blocks in the third eye or sixth light wheel, and opens a channel of communication with the invisible realm. This can be helpful for making contact with plant spirits and fairies as well as generally enhancing clairvoyance, bringing insight and the understanding of dreams."

Whenever one wants to become the hollow bone, or channel information from other dimensional realities or from the spirit of a plant, Elder is the plant to call upon to help open one's channel. Elder offers access to the multiverse through many ways and directions—above and below, inner and outer, and through the elements and other aspects of Nature. During an Elder initiation, Clare experienced Elder's connectivity: "Elder is bringing in new, never-before-seen plants through crystalline structures."

The ultimate veil that we all must step through is that of death, and Elder can serve as our guide. Emma Fitchett (Farrell) in her book *Journeys with Plant Spirits* says:

> We have forgotten how to die in the West, how to treat our dead, and how to ensure all soul fragments are retrieved from all directions before the deceased cross the veil. Death rituals were once adhered to in these lands, and having trained in Celtic funeral rites, I can understand why Elder is such a blessed ally for both the preparation for our own death and our assisting with the death of others. . . . she can give us a taste of death, take us very close to it or even through the veil and bring us back again in order to awaken inner strength and other qualities that we are lacking for our work and life.

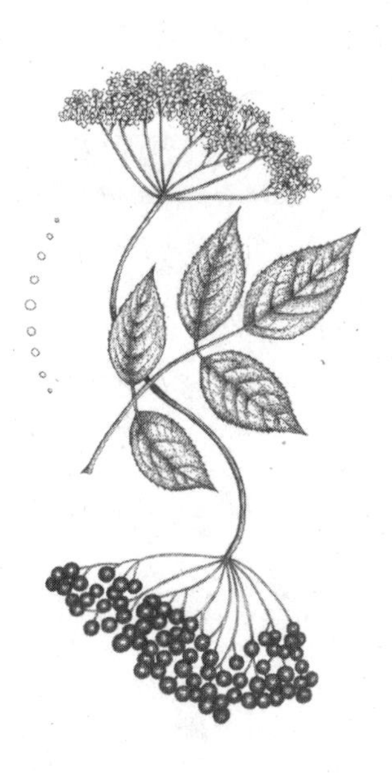

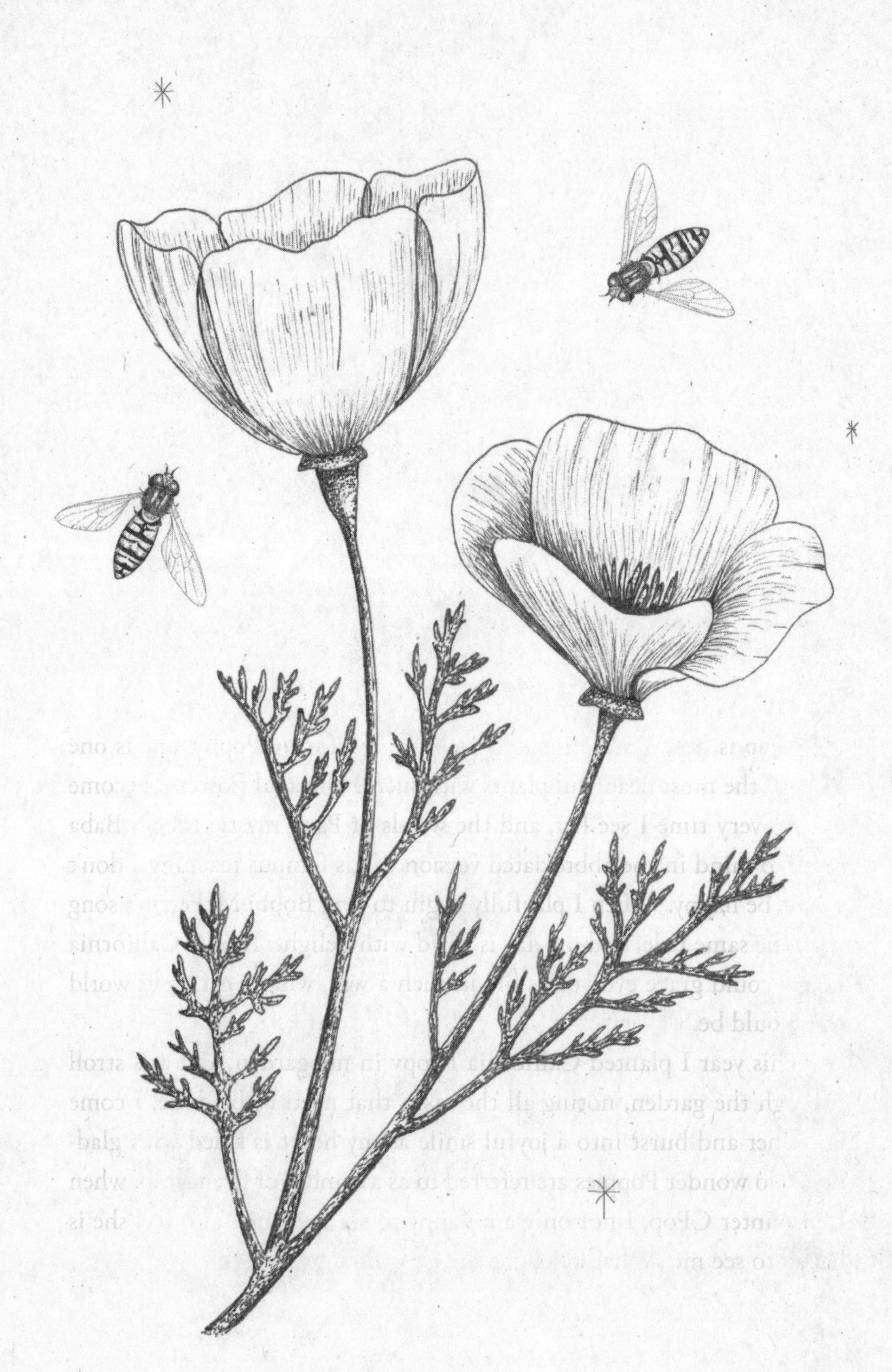

TWENTY-TWO

California Poppy

Eschscholzia californica

CPop is how I affectionately refer to California Poppy. She is one of the most beautiful plants with such a cheerful flower. I become happy every time I see her, and the words of Parsi mystic Meher Baba come to mind in the abbreviated version of his famous teaching, "don't worry, be happy." Then I playfully begin to sing Bobbi McFerrin's song with the same title, and my day is filled with delight. If only California Poppy could grace everyone's life in such a way, what a different world this would be.

This year I planted California Poppy in my garden, and as I stroll through the garden, noting all the tasks that need to be done, I come upon her and burst into a joyful smile as my heart is filled with gladness. No wonder Poppies are referred to as a symbol of friendship; when I encounter CPop, I not only am happy to see her but I also feel she is happy to see me. What a delight she is!

Cup of Gold

This delicate orange or yellow flower with light feathery leaves seems so unassuming, so much so that one would never guess what gifts lie within. While in the dreamscape of CPop, the felt sensation I received was a glow throughout my body, but it seemed solid. The essence of this feeling sensation seemed like what gold might feel like. I referred to it as "cup of gold," which resonated when I vibrated this back to the plant. I traveled into the cup of gold and found a treasure of spiritual gold, which is the currency of California Poppy. In our physical, material world, wealth is measured by the number of objects you have acquired, or property you own, or money you have in the bank, but this kind of wealth comes and goes. Spiritual wealth is not something you can buy, but when you are filled with this kind of gold, you are truly wealthy. There is no limit to this wealth, and the pot of gold never empties. As a currency, spiritual gold does more than buy things; this is what makes you truly happy. When you exchange spiritual gold with others, the return is tenfold in kindness and generosity. When CPop becomes your partner, you are filled with the vibratory resonance of the gold frequency, the frequency of unity consciousness. Because of the deep cavern of spiritual gold that you have access to via California Poppy, you can bestow these riches on all you come in contact with. The touches that you communicate to others are like pouring liquid gold from your inner chalice. When CPop lives inside, you feel her each and every day in your "heart of gold."

As Above, So Below

In many cases, plants have imprints that help you understand what their gifts may be. This impression of the plant is a way of communicating the possibilities that lie within. You can follow the thread of the impression to see if your own empirical evidence supports the suggestion of the imprint. Herbalist Julie Caldwell says, "It's important to note that both the flower and the root of California Poppy embody the gorgeous orange hue. This unique occurrence makes one ponder the significance of this plant signature. Both the flower and the root are receptors, gathering information from the mycelial earth network and the celestial cosmic network bringing them into a singular expression—As above, so below."

The words *as above, so below* often bring to mind the concept of creating heaven on earth, but in reality they originated in an ancient Egyptian Hermetic text called the Emerald Tablet, which had to do with the effect of celestial bodies on earthly events—in other words, astrology. Another explanation of this idea is that the macrocosm is reflected in the microcosm and vice versa; the universe is the macrocosm and humans are the microcosm, with the understanding that we humans have a mini-universe within. Through Julie's experience and observation she was led by California Poppy to understanding her internal universe and how it reflected the bigger picture, as well as, "how best to express our unique creation/life."

Calm Relaxation Brings Transformation

California Poppy has been a component in my sleep formula, which I have taken for years. She creates a state of calm where stress melts away into relaxation. Any anxiety I may be carrying is released and peace prevails. Not only do the chemical constituents of CPop have antianxiety and sedative properties, but the energetics of CPop create what Julie calls "calm coherence," meaning on a spiritual level we are content in our alignment with our life.

What I notice is that when I'm calm and not filled with anxiety, I am more able to cope with life's challenges and focus on solutions instead of problems. I'm able to more clearly see what is needed for my inner transformation, which then is reflected in an outer transformation. During the California Poppy initiation, the transformative insight that came was, "You are already whole and perfect. You just need to remember this original blueprint of you."

One of my students and dear friends, Coco, shared this about her experience with California Poppy.

> *When journeying with Poppy I had the most exquisite experience. I felt Poppy's presence pulsing through my being, creating flow, harmony, and calmness in my being. The energies then started to move down toward my feet, and the threads of light continued to pull me into Earth, through all the layers to its very core. I found myself in an embryonic form in the core of Earth and was floating inside Pachamama herself. Then she slowly birthed me. I began to float back up through the layers and back into my body to feel a deep stillness and my nervous system nourished. This vision had such a profound effect that I continue to this day to visualize my experience to assist me in staying anchored and connected to Pachamama.*

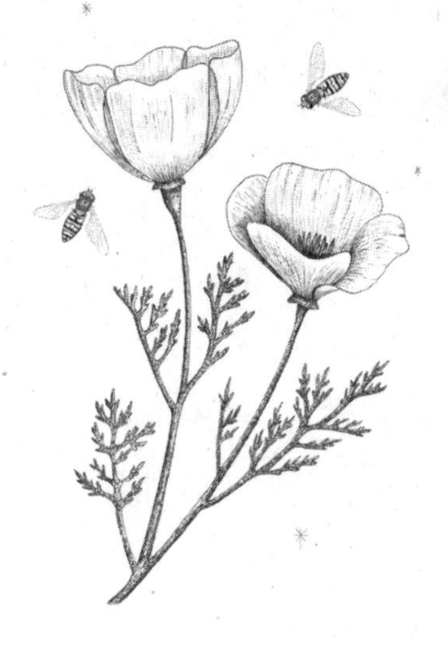

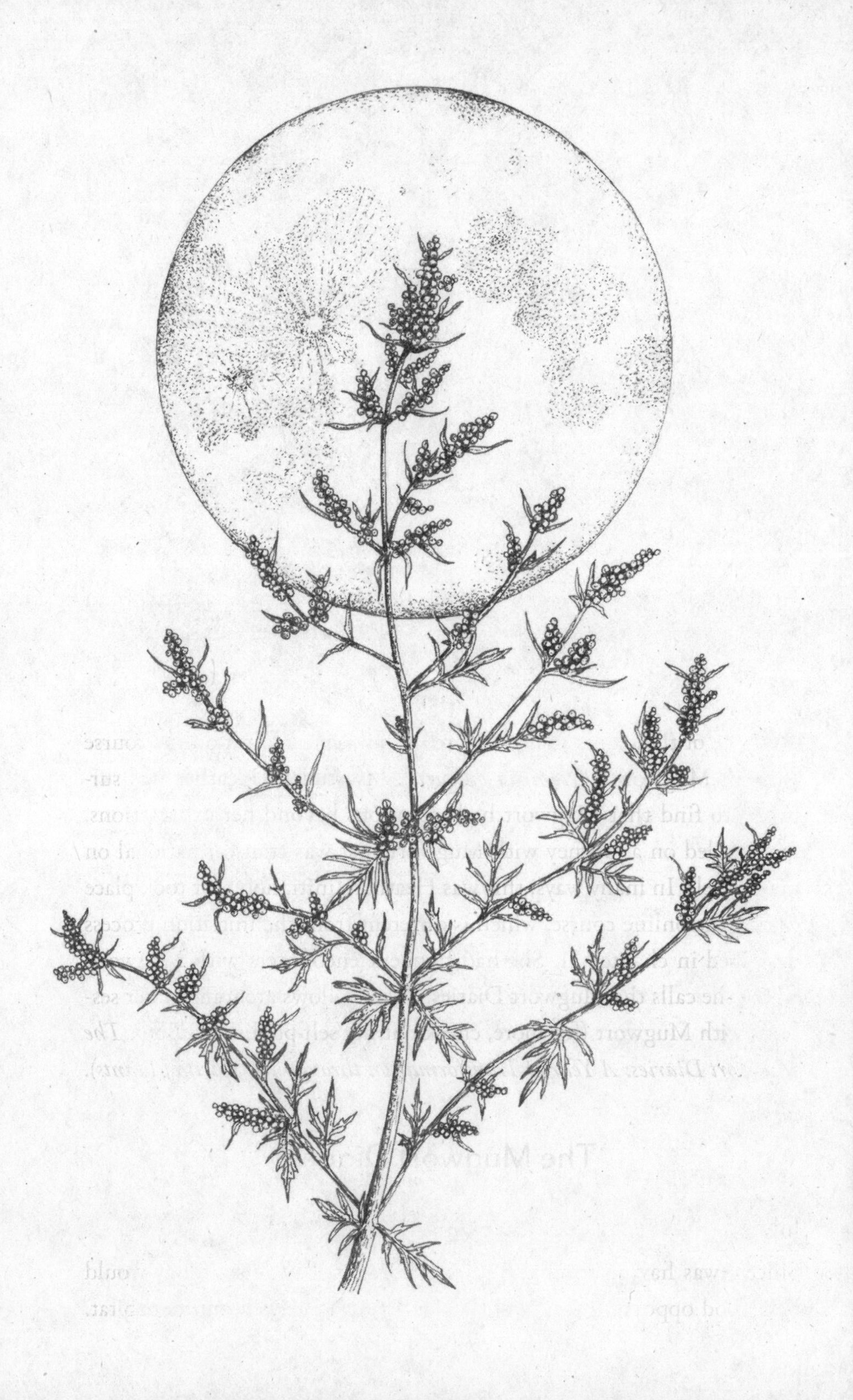

TWENTY-THREE
Mugwort

Artemisia vulgaris

One of the plants that is covered in my eight-month online course is Mugwort (*Artemisia vulgaris*). My student Heather was surprised to find that Mugwort had intentions beyond her expectations. She was led on a journey with Mugwort that was transformational on many levels. In many ways, this was Heather's initiation that took place during the online course, which is different from the initiation process described in chapter 11. She had frequent encounters with Mugwort, which she calls the Mugwort Diaries. What follows are some of her sessions with Mugwort (for more, check out her self-published eBook *The Mugwort Diaries: A Year of Transformation through the Spirit of Plants*).

The Mugwort Diaries

July 8

Since I was having trouble finding Mugwort, I thought today would be a good opportunity to go and look for it. I checked out its habitat.

It grows in coastal areas, so I thought I'd go along the Fife Coast to Wemyss and see if I could see any on a wildflower walk. I saw plenty of plants that I haven't seen for a long time: Century, Agrimony, St John's Wort, Bladder Campion, Teasel, among others. I stopped to look at a Teasel that was on my right-hand side and quite impressive. Thank you, Teasel, for getting my attention, but I need to connect to Mugwort. I turned to my left and looked down, and there she was. I stepped back a few paces, and on my right was another Mugwort plant. This one was much larger than the first Mugwort plant. Now I know where to find them. (Thank you, Teasel, for getting my attention, which was your way of helping me find Mugwort.)

I walked along to the harbor at West Wemyss and looked across to Gardener's Cottage. I hadn't been there for a while. I felt drawn to go along there. I started to walk. I saw one Mugwort plant and then another and another. There were lots of Mugwort plants, both large and small, and I was able to have a close encounter, which helped me to know what Mugwort looks like. I sat beside one large Mugwort plant and introduced myself. A tingly feeling started to creep up my left leg and then it went across my back. The feeling continued for some time before subsiding. When the tingly feeling finished, I collected myself together and said, "I get you." I went home as a different person.

What an afternoon that was! I found Mugwort, though I suspect there was a bit of work in the energetic realm between us to arrange our meeting. I got to introduce myself and spend time with Mugwort. I had a light-bulb moment with Mugwort—and I think an important one—which was that I came to understand how special Mugwort is. It's a bit like meeting someone who is meant to be really special, but you don't really get that about them until you have met them and spent some time with them.

I'm looking forward to spending more time with Mugwort.

July 18

Three years ago I bought a Music of the Plants device from the spiritual community of Damanhur, which, when its electrodes are attached to a part of a plant, perceives the electromagnetic signals the plant is emitting and translates them into musical harmonies. It completely changed how I experienced plants and trees. For the first time I really saw them as living beings and not just inanimate plants or trees. They became beings that reflected on what they were hearing and responded to what was happening around them. They were aware of me and my attention to them. The moments of wiring up a plant and the plant playing its first note or notes are priceless. I love the plants that make a note and then stop to have an aha moment. After reflecting, they continue, and then there is no stopping them as they begin singing. They seem to love to sing, absolutely love it. It is a joy to listen to them developing their abilities over days, weeks, and months.

I was largely working with trees. They taught me to be patient with them. If I was impatient or inattentive, the connection broke up or they stopped. They wanted my full attention and for me to fully listen to them. Each species had its own song, and each tree also seemed to have its song. One Beech tree sounded a little different from another with individual songs. There was even a difference between ones that were good singers and ones that weren't.

One day, while recording one of the famous Cadzow Oaks from the remains of the medieval Cadzow Forest in South Lanarkshire (a county in the Central Lowlands of Scotland), I fell asleep standing on my feet as the tree's music lulled me to sleep for a few minutes. What tree magic was this? The tree was an amazing being: ancient, massive, powerful and a host for other life.

Having made my introduction to the Mugwort plant near my house, I thought it was time to develop our relationship. This particular Mugwort grows in a flower tub placed on the outer wall of the Adambrae Cemetery. It is growing as a "weed," there being no surviving flowers in the tubs as they are no longer looked after. The pavement and road in front are reasonably busy, the former with dog walkers at particular times of the day. Mugwort was willing to engage with me. It was a very windy day, and the plant was a little droopy, as if it needed water. Anyway, I connected Mugwort to the Music of the Plants device, and it wasn't shy or filled with any performance anxiety. In the middle of the session, I looked at Mugwort who seemed to be struggling from a lack of water. I disconnected my device and went home for some water. I gave it four pints and had the sense that it felt a bit better. I reconnected Mugwort to the device.

I became aware of a soft gentle sensation running from my ankle up my left leg. It was strongest on the outside of my leg below my knee. There was a bit on my left shoulder blade. This was like the sensation I experienced when I sat with Mugwort at Wemyss. It flowed up my leg for more than half an hour. The sensation was cool, but pleasantly so. I tuned in to my leg and shoulder blade and then went back to the feeling, which hadn't changed. Soft. Gentle. Cool. It reminded me of the cool feeling of peppermint essence in the bath. I went in and out of the sensation, but it was always the same. Was this the energy of the plant? Mugwort was still singing away as all of this was happening. Its cadence was smooth and steady.

A dog walker went past with a small dog. The dog had a look of horror in its eyes—a look that I won't forget in a hurry. It barked two or three times, and the owner had to pull the distraught dog away. Then the dog looked back as if it was seeing something that I couldn't see. Another small dog passed and looked back at me as if to say, "What on earth?" It eyed me and Mugwort up and down. The passing folk with their dogs must have wondered what was going on, as I was standing against the wall with a device with wires attached to a "weed." I must have looked a bit strange to them. I was a bit uncomfortable standing in such a public space recording Mugwort. It's like going out busking, but with a plant. It occurred to me that I needed to learn about busking (something I have never done) and begin to practice being in this space if I plan to be in very public places communing with plants via the music-producing device.

July 19

I woke up with a stinging pain at the end of my left wrist bone. I had surgery over a week ago to remove a plate to help the broken bone heal.

Reiki usually helped me to calm down and soothe any sensations, but it wasn't noticeably helping. By the time I had my midmorning coffee, my wrist was still stinging. It was going to be a long day. I had intended to go and visit Mugwort to see how she was doing and to help develop our relationship after our music-making session the day before with the Music of the Plants. I headed toward the local cemetery where she was growing. She was looking a lot better after taking up the water that I gave her yesterday. I started to feel a warm smooth feeling in my left foot and up my leg, followed by a growing tingling feeling up my back (my felt sensation of the plant). I yawned (a sign of energy releasing) and settled down as Mugwort began to move through me in a channeling fashion and communicated the following to me.

> *Oh thank you for your care and your continued care. I welcome it. Most people walk past and don't see me even though I am tall.*
>
> *We are connecting differently today. You feel it through your feet. I am connecting you to the ground here to let you see what it is like to make this connection. You feel it in your left foot and that gentle heat—a smooth heat—is busy growing as we speak together. It is getting warmer and warmer. You are thinking this shouldn't be the case, but it is. This is a nice temperature I give you, a reminder to connect with the ground—becoming grounded.*
>
> *When you work with me, you need to connect to the ground as I take you to great heights and places that you would not be able to go to if you aren't grounded. But you need to be on the ground, and I am showing you that this morning. This is not needed much in your right foot because there is a bit of ground in it already, but your left foot needs it. I'm in the ground here, but I have a much greater reach into the air and skies above. There is much magic in this above realm.*
>
> *You are near me and need to be able to sense me. You can't do that at this moment, but I will show you how to sense me. We will work together, and you will see what I can do for you, and you will take what I give you and change the world with it. That is why you are here; I am here to show you how to change and to change the world.*

I am thirsty, but you know that and will give me some water later.

Your foot is still warm as I send heat throughout you. I have brought this heat into you to show you how powerful I can be. Keep on feeling it; the warmth is gentle but powerful. Reflect on this.

My silver color on the back side of my leaf is magic. Look at all the different colors and shapes of me and look and reflect on them. I am giving you a gentle cooling now to balance the heat I have given you. I thank you for coming back to me and we will see you later. We have done well this morning. Thank you.

Soon after I returned home, I became aware that there was no longer a stinging pain in my wrist. My wrist felt settled and calm and has continued to be since my session. The session was very different to the one the day before. It took our relationship forward and established some basic advice and conditions that needed to be put in place before it could be developed further. Mugwort was ensuring that these were in place and that I was fully involved in this. I have some "homework" to complete.

July 21

After the last session with Mugwort, my left wrist had no pain. It felt settled and in a much better place than it had been since surgery.

I was excited to develop my relationship with Mugwort. Today's session felt like an initiation with Mugwort, the plant working on me and through me to prepare me to work with it.

As Mugwort said, "Thank you" at the start of the session, a woman with a child in a pram passed in front of me. The child turned toward me and said, "Thank you."

Oh yes, you are back. Thank you. Thank you for the water, I appreciate it. Yes, thank you. I feel so much better for the water. You could be more careful in how you put it in, but you give me water—and life. I have grown since you last saw me. I know you have noticed it.

You want to work with me. I am happy to do that. You feel that lovely warm feeling in your left foot. Well, that is me. It is just gentle—and up your spine where you have just sensed it and up your back and into your shoulder blade, your left one. I'm all through you and more through your left foot. A warming heat, a gentle heat on this cool morning. You want a word to describe it. Well, how about the warm heat of the summer evenings? Yes, that sounds good, but it is like a different heat, a heat of the noon. It is going up your leg and getting much warmer as you are getting warmer and a bit agitated by it.

I am magic and I catch the magic from the sun and the moon. You could think about that. Or maybe think of something else. I am silver and green and red. I transform when the wind blows through me. How about the silver one in the wind? Yes, silver one in the wind.

You feel that heat in your left foot. You need it. I'm sending it up to you. I said you need to be with the Earth and you do. Just look at that green car—think green and everything green and that will help you to connect with all that is around you.

I move in the breeze. I let it go through me. It ripples my leaves, though not so much today. I am standing taller and straighter today. I can reach out to everything and everyone when I am like this. I can also hide in my branches. They are strong and supportive, and I help those that exist in my branches. Feel that heat. It is gentle, but it has more force in it. You need it. I shall send you some more and keep sending it to you.

Ah, that is the sun out. I do like it on my leaves and flowers just as you like the heat in your left foot. It is rather pleasant. I will let you work out why I am sending this heat into your foot. It will help you in this work that we are doing together.

You ask why have I come to you? Because I have been chosen to work with you and I was sent here to work with you. That work will unfold gradually as we work together. You will need to be patient in this work. All I ask is for some subsistence from you and to recognize me and we can work together for bigger things and to help one another.

I was put here so that you would be able to easily find me. It works well. I will be here for a while to help you.

You still need some heat from me. Just stand and take it in. You are noticing my wonder. Yes, I am wonder. I will keep changing as we work together. Tomorrow I will look different. And the day after that I will look different again. There, that is enough from me today. Look to see what I have changed. It is better for you. You will see that. Thank you.

July 26

After the last session I started to think about silver and the color silver. Around me was silverware in the house—pens, jewelry, candle holders, coins, card holders, parts of pens, mirrors, bathroom fittings, clocks, zips on clothing, my summer cardigan, wrapping paper, and stationary. There was also silvery water and silvery skies. How could I forget silvery garden plants: Silver Cineraria, which I love planting in the garden over the winter months; Snow in Summer, Lamb's Ears, Mullein, and Poplar trees. There was also silver underneath the leaves of Raspberries and Meadowsweet. There was silver everywhere.

Were there any negative energies associated with any of these items? No, silver is always positive, especially when it is part of something. It embellished the bathroom and other rooms. It connected us to essential items and was the interface between us and doing an activity: a fork used to feed ourselves great food; silver handles on the cupboard doors to open to a wonderland of food; taps on the bath that turned on water to provide a lovely relaxing bath. Silver also lets us see ourselves: mirrors serve as our reflectors and can be used to help us transform ourselves by helping us to see when combing hair, putting on lipstick, or deciding on appropriate clothing.

But too much silver was a bit overwhelming. My silver boots that I thought were cool weren't really. That was the only negative thing I could think of. That was the key bit: the balance that Mugwort had spoken of. Even the amount of silver in my life needs to be in balance. With balance we can achieve anything: we can release blockages, we can flow, we can change and transform.

I had a further session with Mugwort:

You can see strength in me today by my strong branches and my tall stems, which are growing by the day. I am coming into my element now as summer is moving along. I know you have been thinking about what I have been saying to you. Yes, you are right: silver can transform everything. I am a plant of transformation, which is possible because of the light that I have, including silver light.

You have just noticed your left foot and the soft heat that is in it. That is me. You know it is me. I am going to be gentle with you today. People look at you as you write. Don't take any notice of them, for you are talking to me and not to them.

Now, where were we? Ah, silver. Yes, it transforms, and I bring transformation. I think you know that your transformation will happen very soon. I was trying to catch your attention earlier today, but you were having none of it. I understand that you have boundaries, but I am also here to help, so embrace me.

You need a handle or name for my vibratory resonance that you experience as warm heat, tingling, and the color silver. Let's work at it again. There. That's me up your spine again. How about that: spine tingles. I give spine tingles and strengthen your spine. Or silver magic? I rather like that. How about silver magic? Yes. You can feel silver magic.

So silver magic it is. And I will be known to you from now on as silver magic.

The passing dog looked at you as the owner called it back. People are starting to notice you now. Just remember to be here now. I have put more warmth up your spine because I thought you needed it. When you think of silver or the color silver, think of me. You can do it a lot of the time. You have identified where silver is used and what it does. So when you see silver or the color silver, think of me, and we will continue to build our relationship together.

I am nearly done for tonight, but just look at how majestic my branches are. Don't they look like candlesticks? Well, here is something else that is silver. I have silver flowers where the flames should be, but my flowers are very powerful, too. I will leave you to think about that. Thank you.

July 28

I have been reflecting on my developing relationship with Mugwort. Something that struck me when she was channeling energy up my spine was that it felt like she was facilitating an initiation for me, like an induction into the tribe. She was preparing me and getting me ready for something, a new state or a new way of being. I thought this at the time, and that this was an odd thing to think about, but now I see that she has been initiating me. She has been telling me about herself and getting me to think about her and what her meaning is and the gifts that she brings. She is consistent in talking about change and transformation that is coming. She is noticing what I need and is providing this through my left foot, while sometimes channeling her energy up my spine. She is providing me with sound practical advice. She is also getting me to think about key ideas and concepts, such as transformation through balance, and today she spoke of flexibility.

Mugwort has also been having a profound healing impact on me. On July 21 I noted that she took the nipping pain out of my left wrist (a plate had been surgically removed from it some eleven days before). My wrist relaxed and has been pain-free ever since. But more than that, I was aware that I felt wonderful, even shortly after surgery. How could I feel so wonderful even a couple of days after anesthetic and getting the plate removed, which had left seven holes to heal? This surgery caused me to become quite unsettled and to experience a lot of localized trauma in my wrist. The doctor that I saw in the postsurgery clinic at my local hospital must have thought that I was mad when I told him how wonderful I felt two weeks after the surgery. He just looked at me with a surprised expression. When I went out of the consulting room, I was struck by seeing everyone in the waiting room slumped in their chairs and looking like life had drained from them. I had thought my joy was caused by having a foreign object removed from me, but all this started before the day of surgery. My sense of being settled, trauma-free, and uplifted started at Wemyss and my first adventures with Mugwort.

What has been slowly dawning on me, especially in the last fortnight as I have been regularly communicating with Mugwort, is that I am returning to being me, the real me. I see me again. I haven't seen me like this for a very long time. It's difficult to explain, but there is this spark that I used to have that went away a number of years ago. I lost it, or the spark became dampened, and I tried to get it back, but it didn't return. Now, I have it back. The best way I can describe the spark is that it's like a deep-seated fire, a passion and enthusiasm. It's having joy. It's being fearless, able to challenge. It's like bringing my whole self into life and everything I do. This is a huge change, and it is all because of Mugwort and the transformation within me that she is helping to facilitate, and this is only the beginning.

I began my wildflower identification work again. I haven't looked at W. Keeble Martin's *The Concise British Flora in Colour* for over forty years. I remember the names of so many of the plants, even after all this time. The plant world is opening up to me again. Out of the blue, a completely "left field" opportunity arose in my rural research activities last week. Is this a coincidence or Mugwort nudging me toward the work that is me and excites me? Needless to say, I accepted the opportunity.

Earlier in the week my cranial sacral therapy (CST) practitioner was a bit bemused when she commented on the energy in my spine. She said that it felt very steady and relaxed in a way that she had not felt before. I have been going to her for sessions since before the start of the COVID-19 pandemic. I hadn't said anything to her about my interactions with Mugwort. I am certain that Mugwort was responsible for the steady energy flow through my spine since she had worked with my spine previous to this appointment.

I had another illuminating session with Mugwort. She continued with her lessons, and this evening the focus was on flexibility and achieving flow in life:

> *You notice that my flowers are becoming more silvery by the day. There are lots of them. I spoke of my strong arms and the flowers that shoot up like flames. Yes, they are my flames. They are all the magic and power that I put out. They are looking splendid.*
>
> *You have been thinking about the things I have been speaking of these last few times we have talked. You are starting to see how powerful I am and how I help to change you. You have just sensed me through your left*

foot. This is me. I am connecting you with you again. Just feel that feeling. Notice it. Sense it. Yes, it is linking you to the ground. I am connecting, grounding, and rooting you to the Earth. Feel the energy as it starts to go up your leg. I will pause to let you notice it. Notice it and not the people along the walkway. Yes. There.

It is breezy. You notice all my silver, all my reflections of duality. This session is about duality in your life and how we all need balance. I am helping to give you balance. You were a bit unbalanced before. You see my stalks and leaves blow in the wind. I balance. I move with the wind. I don't resist. I go back and forth. It is easier than trying to resist. Go with the flow and not against it. If I went against it, I would have broken stems. So that is a lesson for you to go with the flow and respond to life in that way. It brings opportunities, and there are opportunities coming your way if you flow and go with the breeze. Yes, life works out better that way. We can open to many opportunities by letting the energy move freely without resistance.

I am giving you a light touch this evening. I'm going steady as you are starting to balance. But I want to make sure that you return to continue our work together. Even if you are feeling a bit better, that is not the same as feeling great, and this takes time to achieve. We will work at that, and we will continue this process. I think that is it for today. I will see you again very soon. But think of all the things that I have said to you and keep reflecting. Thank you.

July 29

One of the things that I have noticed since I started Pam's course is that plants have been stepping forward into my subconscious realm. Sometimes they fleetingly pass through, just giving me enough time to see them and acknowledge them. After my surgery to get my plate removed from my left wrist, I was dozing in the recovery room and Purple Loosestrife shot across my field of vision. This was quickly followed by Elder. Elder was showing me her strong, straight new branches; I have been working with Elder since earlier in the year. After my first surgery, she had also shown me the image of straight branches. I took it to be a good sign for recovering after a broken wrist. Last night Bramble came forward in my sleep.

Mugwort has also come to me in my subconsciousness in other ways, not from this lifetime but when I was viewing other lifetimes during the last two CST sessions in July. The first session was dealing with the trauma of my broken wrist. It was one of a number of sessions that involved encountering Alexander Orrock, agent to the Wemyss Estates on the Fife Coast, who lived from 1776 to 1836. He had also broken his left wrist and was having treatment from the local bonesetter and the cook at Wemyss Castle. She had provided him with a number of herbal poultices, one of which included Skullcap, to help him get a good night's sleep. Mostly, she did not tell Alexander what herbs she was using. In one of them, she included Mugwort. In the session, he fell asleep and was dreaming with Mugwort: a dream within a dream state. It helped him to rest.

Earlier in the week Mugwort wanted to come into my CST session. Since I am working with her outside the dream state of CST, I pushed her back: she had commented on that on July 26, when she said that "I was trying to catch your attention earlier today, but you were having none of it. I understand that you have boundaries, but I am also here to help, so embrace me." However, I think she was there in another capacity, perhaps to help to heal me in the dream state. There were many dreaming episodes during the CST session: I was a ship commander and I fell into a dream of other places. I fell asleep again and apologized in the narrative for doing so, then I also turned up as myself in this lifetime and fell asleep in the session, and in the dream state I died and went wandering into different realms to explore. I'll be interested to see how this will play out in my physical life in the coming days and weeks.

Mugwort has given me lots of advice to think about. Given all the dream states and reference to dreams, I think I need to start dreaming with her, with a sprig of her under my pillow, and see what guidance she can offer in the dream state.

I had another session with Mugwort:

You have seen that I am growing. I am growing just like you. I have been changing. Yes. You have just noticed me. There. On your left foot. I think I need to give you a bit more today to help you go forward.

There. Get yourself sorted. I'll work on what I need to give you. Yes. I've started through your left foot and slowly up your leg. Feel that heat. Isn't it lovely and gentle? I like doing that for you. And a bit more. There. Feel it. Sense it. Yes, you are feeling it. That is making a difference for you. I will keep going with it. There. Just feel. And now into your right foot. Just keep them still. And feel. No one will disturb you, for I have arranged that. And up your right calf. Feel that it's all getting so much stronger.

You haven't really felt energy like this for a while. I'm making it stronger for you. Warm. And gentle. And flowing. I am keeping it strong. I want to show you how strong I am. There. Keep feeling. Ignore the rain. It might pass. I am adding more power to you now. Feel those extra volts go through you. Keep with it. Isn't that amazing. Keep with it. Focus. Keep focusing.

I'm pulling it up your legs now.

You are thinking of ice cream at Wemyss. One of us gave you a similar session there. You thought it was amazing. Ah, smell that rain. You better go. I will see you later.

The session ended as a result of a heavy downpour of rain.

July 30

We started a really powerful session yesterday but were interrupted by a heavy rain downfall. Mugwort acknowledged the rain and decided that it would be better to end the session and resume at another time.

Writing this diary entry this morning, I have become aware of another shift. In the last few days Mugwort has been talking about flexibility and its importance in going through life. While I could say that I have a fairly flexible approach to what is going on in life, I need to rethink it.

Having a broken left wrist is another way to look at flexibility (I wouldn't recommend breaking a wrist to see how flexibility works) and how to deal with day-to-day things, like getting dressed and all the basics of living such as eating, drinking, and going to the loo. A one-handed existence instead of a two-handed one does toss life in the air and make you relook at it. It is great for learning about strategic planning, but that is another acquired skill. Flexibility also relates to the wrist and being able to use it flexibly and being able to maneuver it to undertake everything with a hand. I'm currently well aware of the amount of stiffness and stuck energy from my left shoulder all the way down to my fingers as my left wrist (and I) are healing from being broken. I am improving day by day, and during this time I'm able to engage more with the world and the normal daily activities that I have probably taken for granted prior to having a broken wrist.

Yesterday after our session, I went out on a walk. I usually find that doing physical therapy exercises with my wrist and hand is so much easier in nature than sitting in the house trying to do them. It is like the plant world around me is supporting me and helping me with my exercises. It happens time and time again. I had a really good session stretching and bending whatever I could.

This morning I've noticed that I have had a big shift in my movement and how I am able to bend my fingers and wrist. I can now scratch the right side of my face. I started to do some activities in an involuntary way with my left hand, including picking up light objects without having to think about it. This all means greater flexibility and less stiffness and stuck energy. Could this be Mugwort's doing? I think yesterday's session with her played a big role in a new level of healing.

After the heavy rain this morning, I had a session with Mugwort. (I have noted in the passage where I was yawning. This was the first time that I have yawned with my local Mugwort plant. I did so in my introductory sessions at Wemyss.)

Yes, I have grown since you last saw me. And you notice my flowers, how much more silver they are—and especially when the sun shines on them, just like now when it has popped out behind the clouds. Yes, we were disturbed yesterday. You will see that it has been blowy, and I have a loose branch. You can take it and use it. I gave you permission to take it. Yes, I am feeling good after all that rain. It was needed. I welcome your water, but it is not the same as the rain. But both are needed. Yes, let's continue from yesterday.

You have already picked up the feeling going through your left foot. That is me. It is increasing in warmth, but it is subtle. Just be with it.

Your tights distract me a bit as they are very bright in color. But I will keep going. There, feel it again. It has cooled a little. Pay attention. There. It is coming. Focus. Yes. There, you feel a tingle at the top of your shoulders. I put that there. You feel that gentle pulse? Yes, you do. Keep with it. I'm giving a flow of energy up your left leg. There is a bit coming into your right foot now. Focus on them. Focus. The breeze is blowing me around a bit. There. That is better.

Feel that up your left leg. I've put a lot through your left foot, and I am building it up. There. That is pleasant for you. Experience it, that gentle feeling. Keep with it. There is a good bit of building. Keep feeling it. [yawn]

You are thinking of Rosslyn Chapel. Why don't you go back? It will support you, too, like I am. Keep focusing. [yawn, yawn, yawn] It is much stronger now. [yawn] Just be with it. [yawn] You are yawning. Continue releasing stuck energy through yawning. [yawn, yawn] You have noticed your left shoulder. Good. I have some energy there. [yawn] Keep focusing. [yawn, yawn, yawn] Feel that shoulder. [yawn] There, more flowing energy is going through you. You really feel it up your lower leg now. [yawn] Keep focusing. Yes, it feels cool, like peppermint cool. [yawn] Your leg feels cool, very cool. Your nose wants to run. Just let it. [yawn, yawn] More. [yawn] There is more across your back. Go on, wipe your nose. That will help you focus. [yawn] I am moving up your left leg. You can feel that. Keep focusing. You are thinking of the biscuits at Rosslyn. You can have one but not just now. Keep focusing. There. There is more going up your lower leg. Keep feeling.

Yes, I have been given a lot of power to change and to challenge. That is what I was sent here for. And I was sent to help you and to bring my powers to you. We will be working together for some time as there is much work to do, and it makes a real change for all of us.

Keep feeling while I am working on you. This is a good skill to have, that of being focused like this and paying attention. It takes some time to learn but it can be done. [sniff] That is me finished with my work for today. I haven't given you much wisdom today as I know that you are still reflecting on what you have learned previously. I am glad that you are reflecting. I will see you again soon. Remember and take that branch of me for you to ponder over and look at. I would be pleased if you did. Thank you.

I took the branch that had been broken off by the wind. It was a strong one from near the ground. I hung it up to dry in my kitchen. I thanked Mugwort for it. I placed the accompanying leaf along with three other ones I had gathered a couple of days ago in my dream purse, which I have now placed under my pillow.

August 1

I mentioned my dream purse in my last diary entry as I wanted to introduce myself to dreaming with Mugwort. I woke up knowing that I had been dreaming and had been in a dream state. I don't remember the dream, but I clearly recollect the advice I was given. It related to a job I needed to undertake in my front garden of laying out a flowerbed. I was given advice on how to do it and the timing. It was a much better suggestion than the one I had intended to put in. I will work with this plan that is very helpful advice.

I wanted to follow up on Mugwort's suggestion of visiting Rosslyn Chapel. As it was a rainy day, I thought this a good opportunity. As usual, I sat on the back pew where there were wonderful views across the chapel and its flower-festooned ceiling. It was an amazing session. There was wave after wave (or cycle after cycle) of energy flowing up my legs and through me. Every time I thought that the session was ending, another wave would come and more energy flowed. There were also plenty of yawns. Owing to the large number of them, I often joke about my "yawn-fest" in the chapel. They got larger and larger as the session progressed. There were some giant ones. My eyes, especially my right eye, were running. There was heat traveling up through me and bursting out onto my forehead. Then there was the cold energy flowing up and through me. At one point I thought I was going to freeze, I was so cold. It was a mammoth session. I was tired afterward.

I met with Mugwort this evening, the first supermoon in August. The clouds were heavy and dark gray, so there was definitely no sighting of the moon. All the cars that passed were either black, white, or gray. There were also two couples with small dogs that walked past. On each occasion, one of the dogs lurched at me and barked, and their owners had to give them a calming word and pull them away. I am glad they were on leashes. It is now routine that dogs who pass are upset by something that is going on with Mugwort and me. Very few walk on without notice. I have again noted in the transcript where I have yawned and sniffed where my nose was runny. The yawning has become a significant element in our work together.

My session today with Mugwort is as follows:

Yes, I am looking great tonight. You have observed that I am growing. I look like pyramids here among my tops. It is good to see you back. I see you have been busy and taken up my advice which is good. I give you plenty of advice. Think about it and act on it. You will see a difference.

Yes, you are noticing that I continue to grow. My flowers will be bursting soon. You are noticing the intricate patterns of my leaves. Isn't it great? I think so too. Go and have a proper look sometime and see what you see in these patterns. What thoughts come to your mind? What do you see? What do you hear? What do they make you think of?

I will get myself ready for our session, so that there is not too much chat. I have started to work with you. You notice that heat coming in your left foot. Yes, you know it is me. I am starting with it—though I hope we don't get stopped by the rain.

Feel that energy. It is very subtle today. You had a lot of energy put through you yesterday by spending time receiving assistance in the chapel. It was good to see you go and have such a good experience. I am different from the chapel, as you know. I am much smaller but have my own strong powers. There. Feel that. Yes. You are picking up on it. Keep with it. Feel it.

You can feel it in your shoulder. I have put it there too. Your shoulder reminds you that there is energy flowing. I am being gentle with you tonight. This is to show you how gentle I can be. Just so that you notice what really gentle is. Feel that. I've increased it a bit and made it more fuzzy and peppermint cool up your lower leg. Notice it. You are feeling it there and up your back. Keep with it. [yawn] Keep with it. [big sneeze] Feel that. Keep feeling it.

You are thinking of a candlestick holder when you look at my branches. I have strong arms to hold my blooms and the power of my flowers. They are like flames of fire but in silver. Keep feeling. That is a bit more. [yawn, yawn]

Feel that. That is a bit more powerful. [yawn] And keep yawning. It is good for you. [yawn]

I could get you to yawn all day. [sniff, yawn] That is peppermint coolness that I am giving you this evening. [yawn] I'm building it up. [sniff, yawn, sniff, yawn, sniff, yawn, yawn, yawn] Feel that up your leg. [yawn, sniff, yawn] I am now showing how much I can send up your leg. [yawn] You know that I can be gentle too. [sniff, yawn, sniff, sniff, yawn, sniff, yawn].

You are feeling warm. Good, I can do that too. [sniff, yawn] I have raised your temperature, and it is moving up you. You can feel it. Keep with it. You do feel warm. [yawn, sniff, yawn] That heat has passed through you. [yawn, yawn, sniff] You are noticing what I am doing with your left foot. Lots of a buzzing feeling. Keep with it. That has rhythm. Keep with it. Keep with it. Good. You are doing well. I'm giving you more up your left leg. Stay with it. A bit more. There. I have given you a lot this evening.

You can dream tonight if you wish. I will try to get you to remember it, and you can reflect on it. You have done well tonight. So keep reflecting on our session, and I will see you soon again. Thank you.

Each session with Mugwort is very different. She is taking me on a journey with her. In the last two sessions with her, she is focusing on her power and energy and getting me to observe it very closely and for me to reflect on it. There has been a significant amount of energy release in these two sessions (as also the one at Rosslyn Chapel). The session this evening had a larger number of yawns, which get larger as the session progresses, and then they subside.

Mugwort Encapsulation

These sessions with Mugwort point out the deep healing that can occur when a plant becomes your ally. The very first thing Mugwort did with Heather was to help her become grounded. A basic law of electricity is that there needs to be a ground wire in order for electricity to flow through an object and, in this case, a person. Mugwort established a ground wire via Heather's feet into the earth. This was needed before Mugwort could begin to move energy through Heather's legs up into her shoulder. The ability to move energy is one of Mugwort's main healing gifts. In my book *Plant Spirit Healing* I share that "the spirit of Mugwort can remove blocked energy, move energy from one place to the next, clear stuck and stagnant energy, and open gates to remove intrusive energy." Mugwort identified that Heather had stuck, stagnant energy that needed to move in order for healing to take place. In the progression of sessions, Heather experienced more movement and release of energy, which had been stuck via her yawning, which became extreme.

Mugwort is also well known for its ability to bring one into alignment, which happened for Heather. When the spine is out of alignment, there is usually a compromise within the integrity of the spine, which can be affected by whether one is living within integrity in their life. Heather's cranial sacral therapy practitioner commented on how her spinal energy felt "steady," which I take to mean not shaky, so her integrity was stable.

One of the signatures of Mugwort (a characteristic marker or form of identification, either physically or energetically) is heat and tingling. When I am working with a client and call on Mugwort to perform an all-in-one healing treatment—my method of asking Mugwort to remove stagnant energy at the base of the skull so that energy can run properly down the spine and bring it into alignment and integrity—I know Mugwort is present and doing her work because I feel heat and tingling in my hands, which are positioned on the back of the client's skull. Heather repeatedly experienced tingling and heat as Mugwort performed healings on her.

Another gift that Mugwort carries is that of facilitating healing and guidance in the dreamtime. Heather was just beginning to touch into the dreamtime with Mugwort when I was no longer able to continue including her accounts of her sessions because the writing of this book was coming to a close.

Clearly, Mugwort was facilitating a deeply transformative process with Heather, which continues to this day. Once an ally, always an ally, and Heather has Mugwort to call upon because they have become beloveds. (***Note:*** When engaging in a Mugwort initiation as I have described in previous chapters, please be aware that it is not recommended for pregnant or breastfeeding women to ingest Mugwort. Mugwort is also not recommended for people who are allergic to celery, fennel, or wild carrot. Instead, an essence of Mugwort can be taken internally.)

EPILOGUE

Choosing a Life-Giving Story

> *You never change things by fighting an existing reality. To change something, build a new model that makes the existing model obsolete.*
>
> Buckminster Fuller

My soul knows my path when I'm born, but there are decisive turning points along this path, and as I live my life, I must choose which fork I will take. My choices might be: becoming one who gives life instead of one who takes life, or being a cooperator instead of a competitor, or experiencing abundance (love) instead of scarcity (fear), or one who embraces unity instead of separation.

When I make choices that are life giving, for not only myself but all of life, I feel my consciousness shift.

I went into the store to get water, and the only option available was to buy a single plastic water bottle. I paused and remembered all the plastic bottles I saw washing up on the beach while near the ocean in Belize. I searched through the drink selection and found one brand of bottled water in glass and chose that one. Of course, I looked at the higher price (it is ingrained in our consumer-based culture to always try to get the best price), but the memory of picking plastic out of the seaweed on that same beach squeezed my heart, and I bought the glass bottle instead. I walked out of the store feeling a wee bit lighter. My conscience didn't weigh quite so heavily today.

My intention to choose life brings my awareness to the state of interbeing, which is part of the story of the New Earth. Could it be that the New Earth already exists—because we have words for it and a vision of it and we are beginning to live this new story? In this Now moment, if I choose life and continue to opt for connection with Nature instead of separation, and I let go of the us-and-them mentality, and I step into unity consciousness and open myself to "the more beautiful world my heart knows is possible" (to quote Charles Eisenstein) and bring this into the next moment, then it becomes my future.

As we begin to think, feel, connect, and act upon interbeing, it becomes not only our present but a part of our past and occupies our memory. As this resonance weaves its way through our collective consciousness, the story changes. The old antiquated way of being becomes archaic, and eventually extinct, because what is emerging is a mass flourishing. Even though we are being told we are in the middle of the sixth mass extinction, is it possible that this is the last gasp on this planet before the Age of Interbeing?

As we step into this never-before-seen physical and spiritual ecology, the energetic template to fully manifest the New Earth emerges.

A Meditation for the New Story

Breathing into your heart, feel your heart soften and move into the radiant brilliance of intuition, unity, unlimited potential, and love. Allow your entire self to marinate in this heart brilliance. Breathing into your gut, feel your belly relaxing into the radiant brilliance of instinctual knowing, primal animal awareness, and truth of Nature. Allow your entire being to immerse in this gut brilliance. Breathing into your mind, revel in the radiant brilliance of mental clarity, creative thought, and deep understanding. As you breathe, allow your brain brilliance to enter the gold frequency where the truth of interbeing is known. As you bathe in this gold frequency, the brilliance of your heart, gut, and mind align and begin to operate with coherent communication throughout your being, helping you to awaken to your true essential nature, the you who knows how to walk a path of beauty that is filled with creativity, abundance, compassion, vision, miracles, generosity, and immense love.

As you breathe into this gold frequency state of being, the old story of separation fades, and the new story that you know is possible begins to emerge. The new story of interbeing—a state of connectedness, interdependence with all of life, and unity consciousness—rises like the phoenix. As you breathe, the vision of this new story becomes brighter, with more clarity, and you realize you're not alone in this vision but are woven into a collective fabric of co-creative partnership with *all* of life. As you breathe, the threads of life are illuminated: plants, trees, water, soil, air, sun, nature spirits, animals, stars, stones, moon, crystals, spirit, love, integrity, truth, potential, abundance, harmony, gratitude, health, unity, peace, cooperation, and interbeing, along with your own original brilliance and the brilliance of others, intertwined into an exquisite tapestry that creates this new story, the story of unity.

As this new story shimmers and each strand becomes ensparkelated, you realize that because you can vision, feel, and love this new story, it already exists in a reality that your heart, soul, and spirit can access. We are on the threshold of a tipping point, when raising our consciousness just a wee bit is all it takes to shift the paradigm into one of unity and interbeing. Together with the help of our Nature kin, we can begin to manifest what we know is possible. Breathing deeply now, let this fresh, vibrant, life-giving story fill you to the brim as you transform into the new you, telling a new story about the New Earth.

Energy cannot be created or destroyed; it can only change form. This means we have all the energy we will ever have at birth, so the question is, what form will your energy take and where will your form of energy be focused? The choice lies completely within ourselves, even though at times it may seem our choices are limited by circumstance. What we are seeing across the globe is that life on Earth is reaching a tipping point. Some see this tipping point through the lens of climate change, while another lens through which to view this moment of transformation is that of the one-hundredth monkey theory, where groups of people across the planet simultaneously engage in consciousness-raising activities. Perhaps the quantum leap of the human species that Myra Jackson speaks of in the foreword of this book is already happening. Together, we are awakening to the New Earth, where living in unity consciousness, as co-creative partners with *all* life, is not only possible but probable.

Acknowledgments

Writing a book seems like a solitary experience, and the initial putting the words on paper occurs when the writer makes it happen. But a book that goes out into the world that you can hold in your hands and read requires lots of folks. In my case I would say "it takes a village."

This book you hold in your hands was a joint effort among many people, but there is one person I am so filled with gratitude for not only being in my life in many ways but for stepping forward to bring her incredible skills in editing to help this book become what it is today. Lauren Valle became my personal editor late in the writing of this book. But she found time within her very busy life (mother of two small children) to bring this book to a point of exquisite readability. Thank you, Lauren, from the bottom of my heart.

Lucinda Warner, artist extraordinaire, brought alive the concept of co-creative partnership with Nature through her stunning art that graces the cover of this book. We talked about the cover and what the possibilities were, but when she sent me the final version I wept. I was so deeply touched by her visual expression because I felt the story I had written was also told on the cover. Besides the cover, Lucinda brought alive each of the plants in the book with her line drawings. Thank you, Lucinda, for your talent, insight, and deep connection with Nature, which made it possible for you to create such beauty.

I first met Myra Jackson when I interviewed her for a Summit I was helping to host. We immediately recognized a kindred spirit in each other, and the following year I invited her to attend the Lady's Mantle initiation here at Sweetwater Sanctuary. Her mystic and visionary gifts shone brightly throughout the initiation, and we came away from that experience knowing we were bonded in an undeniable way. When I began to think of who I would like to write the foreword for this book, Myra instantly came to mind. Thank you, Myra, for becoming my friend and colleague and for writing this inspiring foreword.

The inspiration to write this book came from my involvement with the Organization of Nature Evolutionaries, a nonprofit organization dedicated to working in co-creative partnership with Nature. I want to thank the vison council members Julie, Sara, Lauren, April, and Laura for their commitment to O.N.E. I also want to thank Elyshia Gardner-Holliday, the executive director of O.N.E., for her unwavering devotion to Mother Earth and all her beings and for guiding this organization to educate others about the right for *all* of life to thrive.

This book was in my awareness for years before it actually manifested. It wasn't until I went to Belize the winter of 2023 that I was able to begin writing. The vital life that oozed out of every cell of every plant and animal and out of the ocean and the beautiful people of Belize was so profoundly inspiring I couldn't contain the creative urge any longer. My thanks goes to the Sõhla Project and the folks who have spearheaded the Belize component of this. It was because of you—Shannon, Bruce, Knox, Inza, Tamzen, Hakan, Tilia, and Cypress—that I ended up in Belize. Thank you for the support and friendship you offered me while we all negotiated our way through life in Belize.

There are so many students and friends to thank for contributing stories that made this book relatable. I won't list you all here because the list goes on and on, but I have mentioned you in the book. Thank you for having the courage to tell your story and being willing to share your stories with others.

Thank you to everyone at Inner Traditions International for your guidance and support once the book was in your hands. It's because of all of you that this book is available to the wider world.

Last, but certainly not least, is my dear heart husband, Mark, who deserves so much gratitude. Thank you for your patience, support, and story contribution and for leaving me love notes in front of the coffee pot so I would see them when I woke at 4 a.m. to begin writing. You have inspired me to no end over the years with your own luscious style of writing, your deep-felt connection to the natural world, and your ability to express your gratitude through eloquent and sacred prayer. Thank you from the bottom of my heart.

Thank you dear readers for answering the call to come home to yourselves and to Mother Earth.

Bibliography

Adams, Patch. *Gesundheit!* Rochester, VT: Healing Arts Press, 1998.

Barbiero, Giuseppe, and Rita Berto. "Biophilia as Evolutionary Adaptation: An Onto- and Phylogenetic Framework of Biophilic Design." *Frontiers in Psychology* 12 (July 20, 2021).

Baurick, Tristan. "World's Largest 'Dead Zone' Discovered, and It's Not in the Gulf of Mexico." *Times-Picayune* (New Orleans), May 11, 2018; updated July 12, 2019.

Berry, Thomas. *The Sacred Universe: Earth, Spirituality, and Religion in the Twenty-First Century.* New York: Columbia University Press, 2009.

Black, Leah. "In Nature's Rhythm." Organization of Nature Evolutionaries website, September 2, 2023.

Bohm, David. *Wholeness and the Implicate Order.* London: Routledge, 1980.

Brenner, Eric D., Rainer Stahlberg, Stefano Mancuso, Jorge Vivanco, František Baluška, and Elizabeth Van Volkenburgh. "Plant Neurobiology: An Integrated View of Plant Signaling." *Trends in Plant Science* 11, no. 8 (2006): 413–19.

British Museum. "Who Was Achilles?" (blog post). British Museum website, October 15, 2019.

Brooke Medicine Eagle. "Music." Medicineeagle.com, no date.

Brown, Richard P., and Patricia L. Gerbarg. *The Rhodiola Revolution: Transform Your Health with Herbal Breakthroughs of the 21st Century.* Emmaus, PA: Rodale Press, 2004.

Browning, Elizabeth Barrett. "How Do I Love Thee? (Sonnet 43)." In *Sonnets from Portuguese: A Celebration of Love.* Reprint. New York: St. Martin's Press, 1986. First published 1850.

Buhner, Stephen Harrod. *Earth Grief.* Boulder, CO: Raven Press, 2022.

Bush, Zach. "Finding the Beauty in Yourself." Interview by Lindsey Simcik and Krista Williams. Episode 484, *Almost 30 Podcast*, Los Angeles, December 30, 2021.

———. "Zach Bush, MD's Beautiful Vision for Human & Planetary Evolution." Interview by Rich Roll. *Rich Roll Podcast*, April 24, 2023.

Central Art. "Jukurrpa." Central Art: Aboriginal art store website, no date.

Chamberlain, Lisa. "Maiden, Mother, and Crone: The Wiccan Triple Goddess." Wicca Living website, no date.

Childre, Doc, and Howard Martin. *The HeartMath Solution: The Institute of HeartMath's Revolutionary Program for Engaging the Power of the Heart's Intelligence.* San Francisco, CA: HarperOne, 2000.

Coila, Bridget. "The Influence of Touch on Child Development." *Bridges*, summer 2018.

Eisenstein, Charles. *The Ascent of Humanity: Civilization and the Human Sense of Self.* Berkeley, CA: North Atlantic Books, 2013.

Elkin, A. P. *Aboriginal Men of High Degree: Initiation and Sorcery in the World's Oldest Tradition.* Rochester, VT: Inner Traditions, 1994.

Emoto, Masaru. *The Hidden Messages in Water.* New York: Atria Books, 2005.

Estabrook, Lisa. *Soulflower Plant Spirit Oracle.* Rochester, VT: Findhorn Press, 2022.

Farrell, Emma. *Journeys with Plant Spirits: Plant Consciousness Healing & Natural Magic Practices.* Rochester, VT: Bear and Co., 2021.

Fessler, Leah. "People Who Talk to Pets, Plants, and Cars Are Actually Totally Normal, According to Science." Quartz, March 31, 2017.

Filemyr, Ann. "Healing the Patriarchal Wound." Southwestern College and New Earth Institute website, November 1, 2019.

Fromm, Erich. *The Heart of Man: Its Genius for Good and Evil.* New York: Harper and Row, 1964.

Gagliano, Monica. *Thus Spoke the Plant: A Remarkable Journey of Groundbreaking Scientific Discoveries and Personal Encounters with Plants.* Berkeley, CA: North Atlantic Books, 2018.

Genetic Science Learning Center. "Epigenetics & Inheritance." Genetic Science Learning Center (University of Utah Health Sciences) website, July 15, 2013.

Gladstar, Rosemary. "Growing Awareness with Botanical Sanctuaries." The Science & Art of Herbalism website, no date.

GurujiMa. "Moving Toward Fifth-Dimensional Awareness." Light Omega website, no date.

Guyett, Carole. *Sacred Plant Initiations: Communicating with Plants for Healing and Higher Consciousness*. Rochester, VT: Bear and Co., 2015.

Hahn, Thich Nhat. *The Art of Living:Peace and Freedom in the Here and Now.* San Francisco, CA: Harper One, 2017.

Handwerk, Brian. "An Evolutionary Timeline of Homo Sapiens." *Smithsonian Magazine*, February 2, 2021.

Hare, Brian, and Vanessa Woods. *Survival of the Friendliest: Understanding Our Origins and Rediscovering Our Common Humanity*. New York: Random House, 2020.

Hicks, Esther. *Ask and It is Given, Learning to Manifest Your Desires*. Carlsbad, CA: Hay House, 2004.

Holmes, Heather. *The Mugwort Diaries: A Year of Transformation through the Spirit of Plants*. Self-published, Amazon Digital Services, 2024. Kindle.

Humbach, John A. "Towards a Natural Justice of Right Relationship," in *Human Rights in Philosophy and Practice*, edited by Burton M. Leiser and Tom D. Campbell. New York: Routledge, 2018. First published 2001.

Hunt, Tam. "The Hippies Were Right: It's about Vibrations, Man! A New Theory of Consciousness." *Scientific American*, December 5, 2018.

Hussain, Grace. "Animals Play Crucial Role in Maintaining a Healthy Environment." Sentient website, November 3, 2021.

Indigenous Values Initiative. "The Great Tree of Peace (Skaęhetsiˀkona)." Indigenous Values Initiative website, no date.

Ingerman, Sandra, and Hank Wesselman. *Awakening to the Spirit World: The Shamanic Path of Direct Revelation*. Louisville, CO: Sounds True, 2010.

Jackson, Myra L. "Living Earth." Academia.edu, March 15, 2015.

———. "The Physics of Now: Restoring Our Sacred Bonds with Nature" (webinar). Women Working for the Earth Summit, Organization of Nature Evolutionaries, April 2022.

Kimmerer, Robin Wall. *Braiding Sweetgrass: Indigenous Wisdom, Scientific Knowledge and the Teachings of Plants*. Minneapolis, MN: Milkweed, 2015.

———. "Nature Needs a New Pronoun: To Stop the Age of Extinction, Let's Start by Ditching 'It.'" *Yes!*, March 30, 2015.

King, Anthony. "Evolved to Run—but Not to Exercise." Science section, *Irish Times*, December 3, 2020.

Kröplin, Bernd-Helmut. *The World in a Drop: Memory and Forms of Thought in Water*. Barcelona, Spain: CIMNE, 2005.

Kunze, Valentine. "Part I: A Brief History of Permaculture." Permaculture Collective, February 28, 2018.

Lieberman, Daniel. *Exercised: The Science of Physical Activity, Rest and Health.* New York: Penguin, 2021.

Lind, Martin I., and Foreini Spagopoulou. "Evolutionary Consequences of Epigenetic Inheritance." *Heredity* 121 (2018): 205–9.

Lipton, Bruce. *The Biology of Belief, 10th Anniversary Edition, Unleashing the Power of Consciousness, Matter & Miracles.* Carlsbad, CA: Hay House, 2016.

Lorca, Federico García. "Romance Sonámbulo." In *The Selected Poems of Federico García Lorca*, translated by William Bryant Logan. New York: New Directions, 1955.

Mandel, Dorothy. "Activating the Restorative Response of the Heart." Unpublished, predissertation paper.

Mann, Charles. *1491: New Revelations of the Americas Before Columbus.* New York: Vintage Books, 2006.

Marvi, Mobin, and Majid Ghadiri. "A Mathematical Model for Vibration Behavior Analysis of DNA and Using a Resonant Frequency of DNA for Genome Engineering." *Scientific Reports* 10 (2020).

Mehrabian, Albert. *Nonverbal Communication.* New Brunswick, Canada: Aldine, 1972.

Mehrabian, Albert, and Morton Wiener, "Decoding of Inconsistent Communication." *Journal of Personality and Social Psychology* 6, no. 1 (1967): 109–14.

Migicovsky, Zoë, and Igor Kovalchuk. "Epigenetic Memory in Mammals." *Frontiers in Genetics* 2, no. 28 (June 8, 2011).

Mollison, Bill, and David Holmgren. *Permaculture One: A Perennial Agriculture for Human Settlements.* Sisters Creek, Australia: Tagari, 1979.

Montgomery, Pam. *Partner Earth: A Spiritual Ecology.* Rochester, VT: Destiny Books, 1997.

———. *Plant Spirit Healing: A Guide to Working with Plant Consciousness.* Rochester, VT: Bear and Co., 2008.

Morley, Julie. *Future Sacred: The Connected Creativity of Nature.* Rochester, VT: Park Street Press, 2019.

———. "'Future Sacred: The Connected Creativity of Nature'; an interview with Julie Morley." Interview by Marc Bekoff. *Psychology Today*, February 11, 2021.

Mother Tree Project. "About Mother Trees in the Forest." Mother Tree Project website, no date.

Muehsam, David, and Carlo Ventura. "Life Rhythm as a Symphony of Oscillatory Patterns: Electromagnetic Energy and Sound Vibration

Modulates Gene Expression for Biological Signaling and Healing." *Global Advances in Health and Medicine* 3, no. 2 (March 3, 2014): 40–55.

Nakamura, Jeanne, and Mihaly Csíkszentmihályi. "The Concept of Flow," in *Handbook of Positive Psychology*. edited by C. R. Snyder and Shane J. Lopez. Oxford, UK: Oxford University Press, 2002.

Narby, Jeremy. *The Intelligence of Nature: An Inquiry into Knowledge*. New York: TarcherPerigee, 2006.

NHS (National Health Service). "Overview: Selective serotonin reuptake inhibitors (SSRIs)." National Health Service website, December 8, 2021.

Nouwen, Henri J. M. *You Are the Beloved: 365 Daily Readings and Meditations for Spiritual Living*. Colorado Springs: Convergent Books, 2017.

Peale, Norman Vincent. *The Power of Positive Thinking*. Reprint. New York: Touchstone, 2003.

Popp, Fritz-Albert. "About the Coherence of Biophotons," in *Macroscopic Quantum Coherence: Proceedings of the International Conference*, edited by E. Sassaroli, Y. Srivastava, J. Swain, and A. Widom. Singapore: World Scientific, 1998.

Prechtel, Martín. *The Disobedience of the Daughter of the Sun*. Berkeley, CA: North Atlantic Books, 2005.

———. "The Disobedience of the Daughter of the Sun: A Mayan Tale of Ecstasy, Time, and Finding One's True Form." Interview by Eric LeMay. *New Books Network Podcast*, Northampton, Massachusetts, February 12, 2020.

———. "Saving the Indigenous Soul: An Interview with Martín Prechtel." Interview by Derrick Jensen. *Sun*, April 2001.

———. *Secrets of the Talking Jaguar: Memoirs from the Living Heart of a Mayan Village*. New York: TarcherPerigee, 1999.

Premier Research Labs.. "Biophotons: Humans are 'Beings of Light'" (blog post). Premier Research Labs website, October 19, 2022.

Quinn, Gary. *The Yes Frequency: Master a Positive Belief System and Achieve Mindfulness*. Rochester, VT: Findhorn Press, 2013.

Rama, Swami, Rudolph Ballentine, and Alan Hymes. *Science of Breath: A Practical Guide*. Honesdale, PA: Himalayan Institute Press, 1998.

Robinson, Lawrence, Melinda Smith, Jeanne Segal, and Jennifer Shubin. "The Benefits of Play for Adults." HelpGuide.org, February 2023.

Rushkoff, Douglas. "Evolution Made Us Cooperative, Not Competitive." Team Human, November 6, 2019.

Sandom, Christopher. "Forget Environmental Doom and Gloom: Young People Draw Alternative Vision of Nature's Future." The Conversation, September 12, 2018.

Santi. *The Twelve Spiritual Laws of the Universe: A Pathway to Ascension*. Scotts Valley, CA: CreateSpace, 2011.

Schatzle, Riley. "Live in the Flow: Nikola Tesla: Energy, Frequency & Vibration." bshko (the Brandish Kollective) newsletter, August 10, 2022.

Sheldrake, Rupert. "Morphic Resonance and Morphic Fields: An Introduction." Rupert Sheldrake website, no date.

Simard, Suzanne. *Finding the Mother Tree: Discovering the Wisdom of the Forest*. New York: Knopf Doubleday, 2021.

Sinnott, Edmund. *Cell and Psyche: The Biology of Purpose*. New York: Harper and Brothers, 1961.

Sissons, Terry Herman. *The Big Bang to Now: All of Time in Six Chunks*. Scotts Valley, CA: CreateSpace, 2012.

Solón, Pablo. "The Rights of Mother Earth." Systemic Alternatives website, no date.

Suler, Asia. *Mirrors in the Earth*. Berkeley, CA: North Atlantic Books, 2022.

Tolle, Eckhart. *The Power of Now: A Guide to Spiritual Enlightenment*. Novato, CA: New World Library, 2004.

Turner, Katya. "12 Benefits of Brain and Heart Coherence." Alleviant: Integrated Mental Health website, no date.

Van Zyl, William. "The First Light and Sound in our Universe: Contrasting Evolution and Creation." Five House Publishing, 2021.

von Essen, Carl. *The Hunter's Trance: Nature, Spirit & Ecology*. Hudson, NY: Lindisfarne, 2007.

Walker, Taté. *The Trickster Riots*. Phoenix, AZ: Abalone Mountain, 2022.

Wandersee, James, and Elizabeth Schlussler. "Preventing Plant Blindness." *American Biology Teacher*, 1999.

Ward, Daniel. "In Search of Duende." Language Magazine website, no date.

Weber, Max. *The Protestant Ethic and the Spirit of Capitalism*. London: Merchants Books, 2013.

Wilson, Edward O. *Biophilia*. Cambridge, MA: Harvard University Press, 1984.

Wilson, Edward O., and Stephen R. Kellert, eds. *The Biophilia Hypothesis*. Washington, DC: Island Press, 1995.

Wolf Winters, Carol. "The Feminine Principle: An Evolving Idea." *Quest*, November–December, 2006.

World People's Conference on Climate Change and the Rights of Mother Earth. "Universal Declaration of Rights of Mother Earth." Cochabamba, Bolivia, April 22, 2010.

Yeo, Alyssa. "The Story of Two Wolves." Urban Balance, February 24, 2016.

Index

About the Author

Earth Elder Pam Montgomery has worked and played with herbs and Nature her entire adult life. She passionately embraces her role as a spokesperson for the New Earth while actively engaging as a shifting-paradigm thinker. She has been investigating plants and their intelligent spiritual nature for more than three decades while teaching internationally and virtually on plant spirit healing, spiritual ecology, and co-creative partnership with Nature. Pam embraces her role as a Nature Evolutionary, helping to usher in the Age of Interbeing. She is a founding member of United Plant Savers and more recently the Organization of Nature Evolutionaries, or O.N.E. A percentage of the proceeds from this book will go to the Organization of Nature Evolutionaries.

About the Artist and Cover Art

Lucinda Warner is an herbalist and nature illustrator who lives next to a forest in the South of England. She is fascinated by the relationships that exist within Nature and how healing can arise from these connections. Her love of the natural world has led her to learn about moths, butterflies, birds, and bees, and she loves to draw and paint the plants alongside the creatures that depend on them.

The title of her cover art piece is *Interbeing*, and it represents the connections among all living things. The Tree of Life surrounded by the four elements in the center of the painting symbolizes the fundamental essence of all life on Earth, with the surrounding circles being a few examples of some of the many species that make up the whole. They are connected by golden dots, which signify the biophotons moving between them, representing their connection through a flow of energy. Lucinda chose to represent humankind with a hand to show how our innate creativity is one of the ways we can enter into deep relationship with the world around us. She also wanted to show humans as a part of the web of life rather than outside it or controlling it in some way—no more and no less than the other beings we share our world with.